Antonino Freno and Edmondo Trentin

Hybrid Random Fields

Intelligent Systems Reference Library, Volume 15

Editors-in-Chief

Prof. Janusz Kacprzyk
Systems Research Institute
Polish Academy of Sciences
ul. Newelska 6
01-447 Warsaw
Poland
E-mail: kacprzyk@ibspan.waw.pl

Prof. Lakhmi C. Jain
University of South Australia
Adelaide
Mawson Lakes Campus
South Australia 5095
Australia
E-mail: Lakhmi.jain@unisa.edu.au

Further volumes of this series can be found on our homepage: springer.com

Vol. 1. Christine L. Mumford and Lakhmi C. Jain (Eds.)
Computational Intelligence: Collaboration, Fusion and Emergence, 2009
ISBN 978-3-642-01798-8

Vol. 2. Yuehui Chen and Ajith Abraham
Tree-Structure Based Hybrid Computational Intelligence, 2009
ISBN 978-3-642-04738-1

Vol. 3. Anthony Finn and Steve Scheding
Developments and Challenges for Autonomous Unmanned Vehicles, 2010
ISBN 978-3-642-10703-0

Vol. 4. Lakhmi C. Jain and Chee Peng Lim (Eds.)
Handbook on Decision Making: Techniques and Applications, 2010
ISBN 978-3-642-13638-2

Vol. 5. George A. Anastassiou
Intelligent Mathematics: Computational Analysis, 2010
ISBN 978-3-642-17097-3

Vol. 6. Ludmila Dymowa
Soft Computing in Economics and Finance, 2011
ISBN 978-3-642-17718-7

Vol. 7. Gerasimos G. Rigatos
Modelling and Control for Intelligent Industrial Systems, 2011
ISBN 978-3-642-17874-0

Vol. 8. Edward H.Y. Lim, James N.K. Liu, and Raymond S.T. Lee
Knowledge Seeker – Ontology Modelling for Information Search and Management, 2011
ISBN 978-3-642-17915-0

Vol. 9. Menahem Friedman and Abraham Kandel
Calculus Light, 2011
ISBN 978-3-642-17847-4

Vol. 10. Andreas Tolk and Lakhmi C. Jain
Intelligence-Based Systems Engineering, 2011
ISBN 978-3-642-17930-3

Vol. 11. Samuli Niiranen and Andre Ribeiro (Eds.)
Information Processing and Biological Systems, 2011
ISBN 978-3-642-19620-1

Vol. 12. Florin Gorunescu
Data Mining, 2011
ISBN 978-3-642-19720-8

Vol. 13. Witold Pedrycz and Shyi-Ming Chen (Eds.)
Granular Computing and Intelligent Systems, 2011
ISBN 978-3-642-19819-9

Vol. 14. George A. Anastassiou and Oktay Duman
Towards Intelligent Modeling: Statistical Approximation Theory, 2011
ISBN 978-3-642-19825-0

Vol. 15. Antonino Freno and Edmondo Trentin
Hybrid Random Fields, 2011
ISBN 978-3-642-20307-7

Antonino Freno and Edmondo Trentin

Hybrid Random Fields

A Scalable Approach to Structure and
Parameter Learning in Probabilistic
Graphical Models

 Springer

Dr. Antonino Freno
Università degli Studi di Siena
Dipartimento di Ingegneria dell'Informazione
Via Roma 56
53100 Siena (SI)
Italy
E-mail: freno@dii.unisi.it
http://www.dii.unisi.it/~freno/

Dr. Edmondo Trentin
Università degli Studi di Siena
Dipartimento di Ingegneria dell'Informazione
Via Roma 56
53100 Siena (SI)
Italy
E-mail: trentin@dii.unisi.it
http://www.dii.unisi.it/~trentin/HomePage.html

ISBN 978-3-642-26818-2   ISBN 978-3-642-20308-4 (eBook)

DOI 10.1007/978-3-642-20308-4

Intelligent Systems Reference Library   ISSN 1868-4394

Typeset & Cover Design: Scientific Publishing Services Pvt. Ltd., Chennai, India.

Printed on acid-free paper

9 8 7 6 5 4 3 2 1

springer.com

A Mario Freno, per avermi insegnato la differenza tra essere onesti ed essere fessi; e a Maria Scaramozzino, per avermi insegnato ad apprezzarla.

—nino

Alla mia famiglia

Edmondo

Foreword

The field of graphical models has been growing significantly since the pioneering studies on probabilistic reasoning in intelligent systems. Beginning from Bayesian networks and Markov random fields, this book offers a new perspective deriving from their nice integration, which results in the new framework of hybrid random fields. While reading the book, one early realizes that there is a unifying approach, that seems to be motivated by the question on how existing types of probabilistic graphical models can be properly combined in such a way to obtain model classes that are rich enough to express a wide variety of conditional independence structures. The authors provide evidence that this combination exhibits scalable behavior in parameter and structure learning. This is also supported by massive experimental results to support the claims on the performance of hybrid random fields in different topics, ranging from bioinformatics to information retrieval. The clean integration idea behind hybrid random fields gives rise to the distinguishing feature of the model, namely the dramatic reduction of complexity that opens the doors to large scale real-world problems. The book not only marks an effective direction of investigation with significant experimental advances, but it is also—and perhaps primarily—a guide for the reader through an original trip in the space of probabilistic modeling.

Interestingly, even though the main subject of investigation is quite specific, while digesting the book, one is enriched with a very open view of the field, with full of stimulating connections. The reader finds a vivid presentation, well rooted in the literature, with inspiring historical references. I very much like the philosophical framework to embed scientific issues given at the end of the book. The authors clear the ground for a view of AI as a science that investigates cognitive technologies. While their investigation does not necessarily provide evidence on natural cognition, they offer indeed a number of intriguing insights. Statistical learning methods are labeled as cognitive technologies, rather than cognitive models. Machines equipped with these methods are viewed as tools for extending human cognition to novel domains, thus offering a nice perspective on the somehow ill-posed question on whether or not machines are intelligent.

I was partially involved in some of the advances on hybrid random fields with the authors, but I frankly find that the book goes well beyond the systematic treatment of the subject. Everyone specifically interested in Bayesian networks and Markov random fields should not miss it.

Siena, January 2011   Marco Gori

Acknowledgments

We are profoundly grateful to Manfred Jaeger, whose continuous advice and in-depth reviews of the evolving drafts of the manuscript turned up invaluable to us, and essential to the pursuance of the ultimate fit of this book.

We feel beholden to our friend and advisor Marco Gori, whose far-reaching and keen support to our work is gratefully acknowledged.

We wish to express our gratitude to Lakhmi Jain, Editor of the present Series, who enthusiastically reacted to the proposal of this editorial project, rendering it possible.

Friends and colleagues who reviewed portions of the manuscript according to their respective competences are thankfully credited, namely (in lexicographic order) Giuseppe Bevilacqua, Valerio Biancalana, John Bickle, Stefania Biscetti, Ilaria "Hillary Castles" Castelli, Michael Glodek, Lorenzo Menconi, Sandro Nannini, Duccio Papini, Claudio Saccà, and Giuseppe Varnier.

Siena,   Antonino Freno
January 2011   Edmondo Trentin

Acknowledgments

Contents

Acronyms

ANN	Artificial neural network
BN	Bayesian network
BIC	Bayesian information criterion
cdf	Cumulative distribution function
CPT	Conditional probability table
CRF	Conditional random field
CVLL	Cross-validated log-likelihood
DAG	Directed acyclic graph
DBN	Dynamic Bayesian network
DN	Dependency network
DOA	Degree of agreement
edf	Empirical (cumulative) distribution function
FOL	First-order logic
GMM	Gaussian mixture model
GMRF	Gaussian Markov random field
HMM	Hidden Markov model
HMRF	Hidden Markov random field
HRF	Hybrid random field
iid	Independent and identically distributed
ILP	Inductive logic programming
IR	ItemRank
KBN	Kernel-based Bayesian networks
KHRF	Kernel-based hybrid random field
KMRF	Kernel-based Markov random field
MAP	Maximum *a posteriori*
MB	Markov blanket
MBF	Markov blanket filter
MBM	Markov Blanket Merging
MCMC	Markov chain Monte Carlo
MDL	Minimum description length
ML	Maximum likelihood

MLN	Markov logic network
MMMB	Max-Min Markov Blanket
MRF	Markov random field
MRR	Mean reciprocal rank
NB	Naive Bayes
NPMRF	Nonparanormal Markov random field
NW	Nadaraya-Watson
OOBN	Object oriented Bayesian network
pdf	Probability density function
PW	Parzen window
SR	Success rate
SRL	Statistical relational learning

List of Algorithms

Chapter 1
Introduction

1.1 Manifesto

"Graphical models are a marriage between probability theory and graph theory" (Michael Jordan, 1999 [154]). The basic idea underlying probabilistic graphical models is to offer "a mechanism for exploiting structure in complex distributions to describe them compactly, and in a way that allows them to be constructed and utilized effectively" (Daphne Koller and Nir Friedman, 2009 [174]). They "have their origin in several scientific areas", and "their fundamental and universal applicability is due to a number of factors" (Steffen Lauritzen, 1996 [189]). For example, the generality of graphical models is due to the fact that they "reveal the interrelationships between multiple variables and features of the underlying conditional independence" (Joe Whittaker, 1990 [315]). Moreover, "[c]omplex computations, required to perform inference and learning in sophisticated models, can be expressed in terms of graphical manipulations, in which underlying mathematical expressions are carried along implicitly" (Christopher Bishop, 2006 [30]).

Bayesian networks (BNs) and Markov random fields (MRFs) are far the most popular instances of probabilistic graphical models [237, 167]. They have been attracting the attention of researchers for more than two decades now, and they found a number of applications to a variety of scenarios. Both of them are expression of that intimate, secret marriage between a graph and a probability distribution which Michael Jordan emphasized in his enlightening metaphor. At first glance, BNs and MRFs represent perfect, complementary counterparts of each other. Bayesian networks, indeed, are also known as directed graphical models, since they rely on a directed, acyclic graphical structure. On the other end, Markov random fields are undirected graphical models, meaning that an undirected graph underlies their definition. We will see that the very nature of the respective graphs has a deep connection with the corresponding probabilistic assumptions, namely the capability of the models to represent specific statistical dependence/independence relationships over the random variables that these paradigms picture as vertexes of a graph. It will turn out that the apparent dichotomy between directed and undirected graphical models is justified only partially. As a matter of fact, a significant overlap exists between the

A. Freno and E. Trentin: Hybrid Random Fields, ISRL 15, pp. 1–14.
springerlink.com   © Springer-Verlag Berlin Heidelberg 2011

classes of statistical independences that can be modeled via Bayesian networks and those representable in terms of Markov random fields. In particular, several popular families of learning machines, like the hidden Markov model [251], can be seen as special cases of both paradigms. Nevertheless, the structure of Bayesian networks can ultimately model classes of probability distributions (i.e., families of independence structures) that the undirected graphical structures of Markov random fields cannot fit, and vice-versa. This has been a strong rationale behind a variety of attempts to develop general, unifying frameworks that can cope with the overall class of probabilistic independences at large. Most of the efforts the scientific community made along this direction have lead so far to very generic graphical models, capable of subsuming the classic frameworks, and whose captivating theoretical properties were thoroughly investigated and handed out. Unfortunately, "a definition that's too general is basically vacuous. [...] The engineering challenge is in finding a decent balance between generality and practicality" (Ben Klemens, 2008 [168]). In the realms of machine learning, artificial intelligence, and pattern recognition, this warning addresses—in our humble opinion—those families of machines that either do not have real algorithms (i.e., that are just a mathematical formalism), or whose algorithms cannot be applied to real-world data (e.g., intractability may prevent their application to high-dimensional data).

This book is concerned with the detailed presentation and thorough investigation of a novel probabilistic graphical model, i.e. the *hybrid random field*. Mathematical definitions and proofs of theoretical properties are given in complete detail. In particular, we will see that the new model actually subsumes and extends both Bayesian networks and Markov random fields. Most especially, hybrid random fields are specifically meant to be learning machines, i.e. they come with well-defined training algorithms (both for discrete and continuous-valued domains) which apply to considerable real-world tasks, possibly involving large-scale, high-dimensional data. A special emphasis is hence put on applications, and the experimental evidence illustrated in the book confirms that (*i*) hybrid random fields can be learned much more efficiently than traditional graphical models, scaling up well to high-dimensional datasets, and (*ii*) they yield results whose accuracy is at least as good as that obtained using more traditional paradigms.

The book is intended for a relatively wide audience. Graduate students are likely to benefit from the introduction and survey of the traditional models, as well as from the effort we put into trying to make the overall treatment of the different topics as self-contained as possible. Of course, knowledge of basic concepts of mathematics, statistics, and computer science is required, but most of the critical notions are introduced (informally, at first) throughout the book, or they are reviewed in the appendices. More experienced researchers and scientists would probably direct their scrutiny to the definition of the new model, its properties, and its comparative performance analysis. Practitioners (coping with real-life problems that may fit a graphical modeling perspective) might be especially interested in considering the presented algorithms, and the examples of applications. Finally, scholars with interests rooted in the cognitive sciences will find a cognitive and philosophical reflection on probabilistic graphical models and statistical machine learning in the final part of the book, as well.

The reminder of this chapter offers a qualitative introduction to some preliminary insights in probability and graph theory behind the idea of a probabilistic graphical model (Section 1.2), pointing then out its fundamental concepts and its extensions (Section 1.3). Some of the milestones in the history of the field are summarized in Section 1.4, while Section 1.5 hands out the book overview.

1.2 Statistics, Graphs, and Beyond

For more than half a century, the concepts of random event and its probability have been of the utmost relevance to such areas of computer science as pattern recognition [73], artificial intelligence [269], and machine learning [30]. The data which are fed into a machine and, in turn, the very reaction of the machine itself, may not always be uniquely determined and necessarily deterministic. Under several circumstances, we need to take into account a degree of *uncertainty* in the phenomena under consideration, and in their digital representation in the first place. For instance, a uniquely-defined output may not be computed in cases where the input data are noisy, noise being a random value affecting the information. Under a similar circumstance, the machine is expected to either reject the input (basically, refusing to misbehave), or to come up with the *most likely* response. In the latter case, we need to be aware of the fact that a certain probability of error (i.e., incorrect output) is involved. Again, the machine may be required to adapt to changes in the environment over time, changes that reflect on the hidden laws that govern the underlying 'distribution' of the quantitative features of the data. Adaptation means that the machine moves from a model of (or, hypothesis on) these laws to a renewed one. The new model/hypothesis accounts for the changes in nature, in the light of the observation of some new empirical 'evidence' acquired on the field. Also, nondeterministic computational models have been studied by theoretical computer scientists for decades.

Concerning the aims and scope of the book, the following two examples will be particularly useful for placing them in the proper context. In probabilistic reasoning founded on probabilistic logic, a system may be expected to carry out a process of automatic logical inference on the basis of partial (or noisy) information, according to a specified set of formal deduction rules [237, 9]. These rules are expected to contemplate uncertainty, i.e. to support conclusions that have a definite likelihood which, in turn, is affected by the knowledge (or, belief) of the probability of specific consequences given the corresponding premises.

The second example concerns decision-making machines. In statistical pattern recognition the goal is the development of classifiers that observe an object and assign it to a class (out of a closed number of alternatives), based on a decision rule. The object is represented as a collection of *features* (or, *attributes*), usually the components of a real-valued vector. These attributes are expected to form a compact, characteristic description of the very object in a proper feature space. A feature extraction module is applied in order to capture the attributes from any given object.

The components of the resulting feature vector are thought of as random variables, whose probability distribution depends on the class the observed object belongs to. The decision rule relies on (explicit or implicit) models of such class-conditional distributions. Parametric models [73] assume a known, parametric form for each class-conditional probability density function (pdf), e.g. via mixtures of Gaussian components, and learn the parameters of the pdfs from a training data sample (according to some parametric estimation approach, e.g. the maximum-likelihood technique). Aside from the theoretical limitation of making an arbitrary assumption on the form of the underlying pdf, there are practical problems with this approach. In fact, diagonal covariance matrices are often adopted in the implementations, in order to reduce the complexity of the resulting machine in terms of: (*i*) computational burden; (*ii*) robustness (the lesser the parameters, the higher the generalization capability); and (*iii*) numerical stability (a Gaussian with diagonal covariance matrix reduces to a product of univariate Gaussian pdfs which, if represented in the logarithmic domain, is just a sum of the arguments of the exponentials). We will say more on Gaussian mixture models and density estimation in Chapter 5. On the other hand, non-parametric approaches [278], such as the linear discriminant [210] or the Parzen window [234], do not require the fixed-form parametric assumption. Unfortunately, some of these models are way too simple for capturing the nature of the class-conditional pdfs. For instance, the linear discriminant is mathematically justified only under very special conditions on the probability distributions [73]. Others (e.g., Parzen windows) are very complex, since they rely on the linear combination of as many kernels as the number of training data. Moreover, the Parzen kernels (e.g., standard Gaussian kernels with circular radial basis) entail a local independence assumption over the features. Although the overall combination of such window functions converges to the correct pdf in the asymptotic case (progressively capturing the statistical dependencies as the number of training examples increase), in the limited-sample case the local independence assumption affects the capability of the machine to result in a reliable model of the dependencies at large. Another non-parametric family of machines that learn from data, and that are dramatically popular in the machine learning community, is represented by artificial neural networks (ANNs) [29]. Most neural networks realize models of regression, or density functions, in an implicit manner. The machine is expected to adapt its 'weights' (i.e., real values associated with its neuron-to-neuron connections) in order to develop an internal representation of the correlations among the input variables. In general, though, the ANN is not readily interpretable as an *explicit* model of the pdfs (or, of the statistical dependencies among the variables). This is due to three factors: (*i*) the network architecture tends to hide the properties of the general laws the ANN has learned, encapsulating them within internal representations (that are effective, but hardly readable from an external observer); (*ii*) the network as a whole does not satisfy, in general, the axioms of probability (unless explicit constraints are imposed); (*iii*) the criterion (or objective) function used for training the weights does not focus on probabilities (in most cases, it is somewhat related to the squared distance between the actual network output and the 'target' output a supervisor passes on to the machine during the training process). Since certain relevant families of ANNs can be seen as particular cases of graphical models, we will go back to the issue later on in the book, especially in Chapter 3.

These examples should make it clear that it may not be simple—as it might seem, at first glance—to develop a compact, explicit, and feasible representation of probability distributions (and of the underlying statistical dependencies among the random variables) as the number of the variables and the amount of data become significant. In particular, modeling quantities such as the joint distribution of a given set of n variables, namely $X_1, \ldots, X_n$, by straightforward application of the chain rule for joint probabilities, stating that $P(X_1, \ldots, X_n) = P(X_1)P(X_2 \mid X_1) \ldots P(X_n \mid X_1, \ldots, X_{n-1})$, grows intractable as n increases. Furthermore, it would require an explicit model, estimated from the data, for each one of the probabilistic quantities involved in the product. Now, the question is: can we come up with a truly general solution to these problems?

Before probabilistic graphical models were introduced, there were no special reasons why the answer to our last question should arrive from graph theory [67]. Graphs have long been one of the most crucial data structures in computer science [212]. They are used to represent a variety of relations among individual data. Applications can be found in physics and chemistry (where graphs can be used effectively for modeling molecules), in biology and bioinformatics (e.g., metabolic networks and protein structures), in artificial intelligence (e.g., semantic knowledge and/or relations between entities in natural language processing, in the form of semantic networks), in social sciences (e.g., social networks), World Wide Web-related areas, etc. A number of fundamental algorithms, often NP-complete, have been developed for solving classic problems on graphs (counting, coloring, spanning, matching, finding cliques, finding paths, etc.) [51].

The idea of studying probabilistic properties of graphs traces back to the seminal work by Gilbert, who introduced the notion of random graph in 1959 [112]. An equivalent model—the ER model—was independently proposed at the time by Erdös and Rényi in [81]. In these models, the structure of the graph is considered to be random, meaning that a process is assumed to introduce new edges between pairs of nodes picked up at random, independently and uniformly. This entails a certain pattern of connectivity over the nodes. In recent years, the random graph was extended to the popular construct of random network by Barabási and Albert (among others), whose paradigm is also known as the scale-free (or BA) model [13, 5]. This relies on the assumption that the random graph process is ruled by agents who add new, independent links to the network topology according to a common probability law (eventually leading to a power-law distribution of the graph edges), so that a spontaneous 'preferential attachment' mechanism emerges.

Only a few decades after Gilbert the statistical machine learning community began to exploit the idea of developing novel paradigms that combined probabilistic concepts and graphs in a number of intriguing ways. Since then, several machine learning approaches involving graphs have been proposed and studied in the literature. Although an exhaustive categorization is difficult, four broad, major classes of paradigms can be pointed out:

1. *Probabilistic graphical models* [154, 174], where statistical correlations are
   jointly modeled by means of graphical structures and conditional probability
   distributions involving sets of random variables. This is the family of learning
   machines that this book is about;
2. *Relational learning*, or—more precisely—*inductive logic programming* (ILP)
   [192], where symbolic representations based on first-order logic are derived from
   examples and background knowledge;
3. *Statistical relational learning* [109], which is a research area focusing on systems that integrate relational formalisms (such as first-order logic) and statistical
   models (such as probabilistic graphical models). A brief survey of this research
   field will be offered in Chapter 8;
4. *Machine learning over structured domains* (i.e. over feature spaces that involve
   a graphical representation [125], where (roughly speaking) each datum is represented by a graph, and an automatic learner is expected to compute functions
   over graphical inputs. Kernel machines with kernels defined over graphs are an
   example [163, 105, 44]. Other instances can be found in the neural networks
   community [308]. *Ad hoc* neural network architectures and training algorithms
   were proposed and thoroughly investigated, including recursive neural networks
   [285], graph neural networks [116], and other models [309].

In summary, graphical models realize the tight binding between probability and
graph theory by representing random variables as the nodes of a graph, and statistical (in)dependencies among the variables by means of the graph edges. The
definition is completed introducing suitable models of the conditional probability
distribution of each variable given (a subset of) the others. A graphical model is expected to yield an overall representation of the joint (or, of a conditional) probability
distribution defined over the whole set of variables. As a distinguishing feature, the
expression 'graphical model' implies that *the machine itself is a graph*.

1.3 Probabilistic Graphical Models

The most straightforward manner of using a graphical structure for representing a
joint probability distribution $P(X_1, \ldots, X_n)$ over n random variables would be to use
a complete (i.e. fully connected) graph having n nodes, one for each variable, and
edges between any pairs of vertices representing all possible conditional dependencies among the variables. Models for expressing the conditional distributions would
be attached (as 'labels') to the vertices, somehow encapsulating the individual terms
involved in the factorization of the joint density, in the form of a product, according
to the chain rule. Albeit feasible, this perspective would result of limited interest,
both from the conceptual and the computational viewpoints. In fact, it would turn
up to be a mere data structure, not offering any advantages over the raw probabilistic definition of the problem. On the contrary, graphical models are expected
to yield nice properties in terms of expressiveness, and advantages in terms of the

complexity of computing the joint distribution. First of all, we want them to exhibit someway a sparse representation of the connectivity between variables, such that an edge is introduced in the structure only if it captures a relevant, direct statistical correlation. Then, we want the graphical model to rely on a certain principle of locality, meaning that the overall joint density can be factorized over conditional probabilities that are defined over 'neighborhoods' of individual variables. At the same time, of course, the resulting factorization shall not only be easier and/or faster to compute, but it must be consistent with the real probability distribution. This implies that either the relevant statistical correlations hold only at the local neighborhoods level (that is, distant variables are independent of each other) or, in the general case, that non-neighboring variables can be thought of as affecting each other only as a side-effect of influencing intermediate, neighboring variables. This requirement can be seen as an instance of the popular Markov property.

One point we would stress is that, in the perspective taken in the book (which is rooted in computer science and engineering), graphical models are not (or should not be) just a powerful formalism with nice theoretical properties. In the first place, they are *useful* learning (and reasoning) machines. As we say, this means that, aside from inference, each instance of a graphical model shall come with effective *learning* algorithms. To this end, the meaning of the word 'learning' is twofold. On the one hand, the graphical structure has to be inferred from a collection of examples (the training data set) collected on the field. Relying on structure learning techniques, the model is expected to develop a topology which best captures the (in)dependencies among the variables, such that it fits well the training data according to some evaluation criterion which must be mathematically sound. Although the graph may be a viable representation of prior knowledge on the application domain, easing the communication of results among scientists, in the general case the statistical relations among the variables are not known (or they are known only partially) in advance. *Structure learning* methods are thence sought that can discover the relations automatically. It goes without saying that, since the set of possible graphs for a fixed number of nodes is countable and finite, in principle the optimal structure could be discovered by a direct, exhaustive evaluation of all alternatives. As a matter of fact, it is seen that this would be infeasible, for at least one (obvious) reason. In fact, in terms of computational complexity, the approach is utterly intractable as the number of variables increases. This problem is clearly taken into account in the development of suitable methodologies for structure learning. On the other hand, once a structure is given (either known or learned), algorithms for learning the local (conditional) probability distributions associated with the variables are needed. Their task is usually referred to as *parameter learning*. The expression implicitly depicts the graphical model as a parametric machine, assuming a fixed and known form for the expression of the probability densities involved in the estimation of the overall joint distribution. Thus, the densities are uniquely determined by the values of a certain set of parameters (like in the case of a Gaussian density, which is fully known once we fix specific values for its parameters μ and σ^2). Algorithms for parameter learning, similarly to structure learning procedures, need to be computationally feasible and robust. We will see that, basically, both of them focus on the same evaluation criterion, stemming out of the maximization of a function related to the likelihood of the model given the data. All of the families of graphical models presented in the following chapters will be provided with an account and analysis of suitable algorithms for structure and parameter learning. The behavior of the algorithms will be evaluated on the field, as well.

The two most popular families of graphical models are reviewed in detail in the book, namely Bayesian networks and Markov random fields (Chapter 2 and Chapter 3, respectively). Since the novel hybrid random field we introduce in the following chapters builds on these traditional models, their thorough description and understanding will be necessary. Several extensions to the models were proposed in the literature. Some of the most significant will be reviewed concisely as well, including dynamic Bayesian networks and hidden Markov models (which extend the basic BN model to time series), conditional random fields (which are a discriminative version of Markov random fields, also suitable for sequence modeling), and some families of neural networks (Hopfield networks and Boltzmann machines, thought of as particular cases of MRFs). Dependency networks [128] and graphical chain models (also known as chain graphs) [191, 314] are also importantly related to hybrid random fields. Such relationships are briefly discussed in Chapter 4.

Other types of graphical models that are not covered in the book are factor graphs [180], clique trees [155], and ancestral graph Markov models [257]. Factor graphs are undirected, bipartite graphs, or hyper-graphs, having two types of nodes, namely the variable nodes and the function ('factor') nodes. It is assumed that a function defined over the whole set of variables (for instance, their joint probability) can be factorized into the product of the factor functions. Each factor, in turn, is defined over a subset of the variables. Edges connect each factor vertex with all the variable nodes that form the definition domain of the corresponding factor function. Factor graphs can be used for the efficient computation of the marginal distribution of individual variables given the others, by means of the sum-product algorithm [180]. Thereafter, factor graphs may be viable models for carrying out Bayesian inference over a set of random variables. Clique trees, also known as junction trees, are the transformation in the form of trees of generic graphs (roughly speaking, cycles are removed via the introduction of meta-nodes that group up and replace the original, cyclic subgraphs). Used along with the tree-junction algorithm [155], clique trees are an effective way of accomplishing exact marginalization (i.e., Bayesian inference) over a generic graphical structure. Finally, an ancestral graph Markov model represents random variables as the nodes of an ancestral graph, that is a structure composed of a subgraph whose edges are directed (like in a BN), another subgraph which is undirected (like in a MRF), and a set of bidirected edges that basically connect the two subgraphs. The resulting independence model, which is closed under marginalization and conditioning, can be shown to contain all independence models having the form of a directed acyclic graph.

1.4 A Piece of History

Implicitly, early contributions to the field range from the seminal work by Thomas Bayes (a Presbyterian minister), published in 1763 [17], which constitutes the basis for Bayesian statistics and the germinal form of his homonymic theorem (see Appendix A), through the research conducted in 1906 by Andrey Andreyevich Markov (a mathematician excommunicated from the Russian Orthodox Church) on stochastic processes and the conditional independence property named after him [207], to the generalization of Markov chains proposed by Andrey Nikolaevich Kolmogorov in 1936 [178].

In 1902, in his book *Elementary Principles of Statistical Mechanics* [111], J. Willard Gibbs provided the scientific community with a model of interacting particles, thought of as being located at given sites within a substance and being in a certain state. The interactions among particles are assumed to be mostly local, within a certain neighborhood of each particle, and to affect the state of neighboring particles. The local energy produced by the interactions contributes to the overall energy of the system. In turn, the probability of the system being in a certain state is expressed, in terms of the Gibbs distribution, as a function of its energy. A similar expression is found in Markov random fields, as we will see in Chapter 3, where MRFs are informally introduced by referring to another example rooted in physics (and strictly related to Gibbs' model), namely the Ising model, developed in the early Twenties [194].

In 1913, John Henry Wigmore (jurist, and expert in 'law of evidence') proposed a graphical method, since then known as the Wigmore chart, which was conceived as a tool for analyzing the flow of legal evidence in trials [316]. Representation of the data and principles of inference in Wigmorean analysis can be seen as the vanguards of modern Bayesian networks.

Path analysis, nowadays widely used in such areas as statistics, World Wide Web applications, and the social sciences, was originally proposed and investigated in biological sciences by geneticist Sewall Green Wright, in the early Twenties, in his famous paper "Correlation and Causation" [318]. Wright developed population genetics, relying on a biological model that is basically a rough prototype of causal Bayesian networks, that contributed establishing the fundamentals of the *modern synthesis* between evolution theory and genetics. In Wright's path analysis the model can be represented as a directed graph encapsulating inheritance, mutations and genetic drifts in populations of animal species.

As reported in [189], also the work (dated 1935) by Maurice Stevenson Bartlett on contingency tables interaction [15] has to be considered a milestone in the historical path toward the establishment of probabilistic graphical models, due to its overlapping with the equivalent notion of interaction in statistical mechanics.

The formalization of the construct of probabilistic graphical model, with its contemporary meaning in statistics and machine learning, can be traced back to the year 1974 with John Moussouris' paper on Markov random systems under the Gibbs distribution [221]. Markov random fields received their quintessential ratification in 1980, thanks to Ross Kindermann and James Laurie Snell's work [167]. A lustrum later Bayesian networks followed, coined in one paper of Judea Pearl's, published in 1985 [235]. Their mature systematization was brought about by Pearl in 1988, with the book *Probabilistic Reasoning in Intelligent Systems* [237], and Richard Neapolitan in 1990, with the book *Probabilistic Reasoning in Expert Systems* [225]. Extensions and variants of these classic graphical models (dynamic Bayesian networks, conditional random fields, etc.) followed them up, as referenced in Section 1.3 and in the following chapters.

Excellent surveys on the topic can be found in [265, 47, 189, 154, 226, 152]. To the best of our knowledge, the most comprehensive textbook covering both learning and reasoning techniques for graphical models is *Probabilistic Graphical Models* by Daphne Koller and Nir Friedman, published in 2009 [174]. Of course, this short list of introductory texts is by no means intended to be exhaustive, given the considerable amount of high-quality presentations that can be found on the subject. Concerning instead the latest developments in the area of graphical models, some of the most representative results can be found—just to mention a few papers—in [6, 171, 115, 323, 241, 326, 200, 277, 182].

1.5 Overview of the Book

In the next two chapters we review in detail the most relevant families of graphical models, namely Bayesian networks (Chapter 2) and Markov random fields (Chapter 3). Thorough understanding of these paradigms (including the underlying mathematical concepts, as well as the algorithms) is necessary, since they can be seen as the building blocks we are later going to use in the definition and development of the novel hybrid framework. The subsequent three chapters form the core of the book. First, they introduce the hybrid random field model (Chapter 4), providing the reader with its formal definition, its theoretical properties, and its learning algorithms. Second, they extend the paradigm to continuous-valued variables (Chapter 5), outlining a variety of techniques for estimating conditional density functions defined over real-valued random vectors. Several applications of graphical models are then presented (Chapter 6), including an evaluation on the field of the proposed model, which is compared to the traditional paradigms. Finally, a concise philosophical reflection on graphical models (from a cognitive science standpoint) is given in Chapter 7. Here's a complete outline of the book:

- **Chapter 2** introduces the idea of Bayesian network (a.k.a. belief network) in a qualitative manner, through the notions of Bayesian inference and causality. A formal definition is then given, stressing concepts that will play a major role in the definition of hybrid random fields: Markov condition, d-separation, Markov blanket, and their relationship with the factorization of the overall, joint distribution of the variables represented by the whole network. Algorithms for learning both the parameters and the structure (i.e., the conditional independences among the variables) of the model from a training data sample are presented. The naive Bayes classifier is then discussed, showing that it can be seen (and justified) as a particular instance of Bayesian network. The chapter is concluded by a summary, a survey of generalizations of the basic model (such as dynamic Bayesian networks and hidden Markov models), a selected sample of real-world applications of Bayes nets in several areas of interest, and a list of the main points of strength and limitations of the paradigm;

- **Chapter 3** makes a step-by-step introduction to the major ideas behind the concept of Markov random field (a.k.a. Markov network), starting from stochastic processes, then moving to the general notion of random 'field' (in mathematics, over topological spaces; and in physics, by referring to the Ising model). The role of potential energy of local neighborhoods over a lattice (an undirected graph) is pointed out, along with the motivations behind the choice of modeling the overall joint density via the Gibbs distribution. To this end, once the formal definition of the model is given (involving maximal cliques, feature functions, and potential functions), the Hammersley-Clifford theorem is stated, handing out the theoretical justification for the definition of Markov random field in the light of the equivalence between the assumption of the Gibbs distribution and the emergence of the Markov property (at the local and global level). It will be shown that the computation of the overall likelihood of the model given the data is intractable. This provides us with the rationale for resorting to a simplified, yet well-behaved alternative measure, namely Besag's pseudo-likelihood (relying on Markov blankets). Algorithms aimed at the maximization of the pseudo-likelihood criterion are then outlined for learning the MRF parameters (that is, the weights of the potential functions, via gradient-based methods) and its structure (via an efficient, regularized search strategy in the space of possible graphs). The chapter concludes with a concise summary, an overview of significant variants of Markov networks (e.g., conditional random fields, Hopfield networks, Boltzmann machines), a synthesis of some representative applications, and a consideration of advantages and drawbacks of the Markov network model;
- **Chapter 4** introduces the hybrid random field model (HRF), in the discrete variables setup. Motivations for the development of the new paradigm are offered first, relying on the analysis of the limitation of BNs and MRFs outlined in the previous chapters, as well as on the crucial point of the computational complexity of learning algorithms for datasets involving a large number of variables (that is, the problem of *scalability*). The model—which is expected to yield a suitable representation of the joint probability distribution of its variables—is then defined, in mathematical terms, as a collection of Bayesian networks over a set of random variables (one Bayesian network for each one of the variables) that satisfies certain conditions involving the concepts of *directed* and *undirected union* (over graphs). Eventually, the definition can be shown to entail (as a theorem) that the model turns out to be a well-defined representation of a joint probability distribution, and that this joint distribution can be recovered (via Gibbs sampling) from the HRF 'modules', namely the conditional distributions of each variable given its Markov blanket in the corresponding BN. The chapter discusses also the difference between this definition of HRF and a looser definition of the paradigm that was presented earlier in the literature (which assumed the modularity property as an explicit part of the definition). We then give theorems proving that

the class of independence structures representable via hybrid random fields is larger than the corresponding classes modeled via either Bayesian networks or Markov random fields. More precisely, the former turns out to be the union of the latter ones. Aside from theoretical properties, our main concern is the scalability of (learning techniques for) the model, that is its capability of being applied effectively to tasks having real-life scale. In this respect, the chapter introduces techniques for inference (relying on ordered Gibbs sampling), parameter learning (a straightforward replica of parameter learning in BNs, iterated over the small modules composing the HRF), and structure learning (relying on a fast, approximate algorithm called 'Markov Blanket Merging'). Complexity issues are then discussed, showing that the resulting learning machine is much more efficient than traditional paradigms. The major differences with respect to other related approaches (e.g., dependency networks and graphical chain models) are finally evaluated;

- **Chapter 5** generalizes hybrid random fields to domains involving continuous variables. Although a large part of the literature on graphical models assumes discrete random variables, several relevant and difficult problems involve real-valued random vectors. This is the case for most applications of statistical pattern recognition and machine learning. The chapter gives the reader a perspective on this scenario by stressing the urgency of dealing with real-valued domains. It then points out that the definition and the formal properties of HRFs do not change, in principle, in the continuous framework, provided that proper models and estimation techniques are given for modeling the conditional pdfs associated with the nodes of the graphs. Fundamental concepts underlying parametric estimation techniques are then reported, leading to the notion of parametric HRF. An emphasis is put on Normal distributions and Gaussian mixture models. Maximum-likelihood estimation of the parameters of (mixtures of) Normal densities is reviewed in detail. *Semiparametric hybrid random fields* are introduced next. They rely on a mathematical result, known as the 'change of variables' theorem, that allows for the development of a *nonparanormal* (or *nonparametric Normal*) technique for the estimation of joint and conditional pdfs without making any prior assumption on the form of the modeled density functions. An alternative framework, the *nonparametric hybrid random field*, is presented eventually, relying on kernel-based estimation of conditional densities (along the lines of the classic Parzen window method). Effective application of kernel-based density estimators requires to cope with the problem of bandwidth selection. A viable solution to this model selection problem is offered in the form of a double-dichotomic search algorithm, based on the cross-validated log-likelihood (CVLL) criterion. Since the complexity of computing the CVLL function is quadratic, an efficient and mathematically sound approximation is introduced via the dual-tree recursion method (based on a Monte Carlo technique). Note that these techniques are suitable for parameter learning in hybrid random fields and in other graphical models, as well. Therefore, the reader may usefully regard the presented density estimation methods also as a complement to the material covered in Chapters 2–3. Finally, Chapter 5 shifts its focus to structure learning in the continuum setting.

It will be seen that the Markov Blanket Merging algorithm developed for discrete HRFs may still be applied, with slight modifications, provided that a viable statistical test is used for determining the correlation between pairs of variables (i.e., during the model initialization phase), and that suitably revised heuristic functions are adopted;

- **Chapter 6** puts the new model at work. As we say, we are concerned with learning machines that effectively learn and prove to be useful in real-world scenarios. This chapter describes some applications of hybrid random fields, with systematic experimental comparison to more traditional (graphical) models and approaches. A general feature selection technique, based on the identification of suitable Markov blankets, is introduced first. Experiments are then carried out on several datasets, including e.g. the identification of boundaries between introns and exons in human DNA sequences. Several pattern recognition tasks (over discrete and continuous domains) are then considered, including the discrimination between positive and negative individuals in a lung cancer diagnosis application. A link-prediction application of the utmost relevance to the World Wide Web is also presented, showing how graphical models can be used for predicting/suggesting references in scientific papers, or for developing recommender systems over a movie database. These tasks are formalized in terms of a ranking strategy relying on a probabilistic scoring function realized by graphical models. Presentation of the experiments includes an analysis of the behavior of HRFs, and a comparative evaluation of the algorithms for parameter and structure learning both in terms of prediction accuracy and computational burden. The results bring empirical evidence in support of the central point of this book, i.e., that hybrid random fields scale up well to large datasets, displaying a level of accuracy which is (on average) at least as good as that achieved by traditional graphical models, along with significant improvements in terms of learning efficiency;

- **Chapter 7** offers a philosophical analysis of some issues concerning the relationships between statistical machine learning (and, in particular, research on probabilistic graphical models) and the cognitive sciences [18, 19]. First, a couple of arguments is put forward in order to correctly locate artificial intelligence—of which machine learning is regarded as a branch—within the framework of the cognitive and behavioral sciences [22, 159]. In this respect, we argue that a correct philosophical interpretation of machine learning and AI should avoid the misguiding (albeit widespread) conception according to which the ultimate success of AI research depends on whether natural cognition can be successfully explained in terms of computational processes. Then, we suggest that a fruitful shift in the philosophy of AI would consist in moving from a cognitive science perspective to a *cognitive technology* perspective, or—in even more philosophical terms—from a view of machine learning and AI as 'descriptive' sciences to a conception of them as 'normative' sciences. A paradigmatic application of this perspective to a couple of philosophical problems—involving the role of simplicity in theory choice and the traditional class of cognitive 'virtues'—is finally offered, with an explicit focus on the statistical machine learning framework underlying this book;

- **Chapter 8** hands out the final remarks and draws the main conclusions, outlining some intriguing directions for further research within the proposed paradigm;
- **Appendix A** and **Appendix B** state the fundamental definitions, results, and properties in probability and graph theory respectively, which are systematically exploited and referred to throughout the book. Although such appendices are not meant to be exhaustive, given the breadth and complexity of the involved fields, they are instead expected to make the book as self-contained as possible, also to less experienced readers.

Chapter 2
Bayesian Networks

"Tho' distant objects may sometimes seem productive of each
other, they are commonly found upon examination to be link'd
by a chain of causes, which are contiguous among themselves,
and to the distant objects; and when in any particular instance
we cannot discover this connexion, we still presume it to exist."

David Hume, 1739 [147]

2.1 Introduction

For some mysterious reasons, Bayesian networks are often introduced in the literature with examples concerning weather forecast (e.g., 'Is it likely to be sunny on Sunday given the fact that it is raining on Saturday, and that my granny's back hurts?'), or gardening issues (e.g., 'How plausible is it that my courtyard is wet given the fact that it is raining and that the sprinkler is on?'). This may give the reader the somewhat embarrassing idea that Bayesian networks are a sort of tool you may be wishing to buy at the "Lawn & Outdoor" level of the shopping mall at the corner. As a matter of fact, there are more intriguing, yet useful, domains: for instance, soccer games[1]. Suppose you want (anybody wants) to make a prediction on whether your favorite soccer team will win, lose, or draw tonight International League game. You can suitably describe the outcome of the match as a random variable Y, possibly taking any one of the three different values w (win), l (lose), or d (draw), according to a certain (yet, unknown) probability distribution. As unknown as this distribution may be, some probabilistic reasoning, and your prior knowledge of the situation, may help you out predicting the final outcome. First of all, you might rely on your estimate of a certain prior probability $P(Y)$ of either w, l, or d (if you do not possess such a prior, you are definitely not a soccer fan). This estimate may involve statistical evaluations on expected competitiveness of the two teams, their current domestic and international rankings, the recent history of their performance, etc. A reasonable estimate of $P(Y)$ may lead to a likely prediction of the outcome, yet involving an ineluctable risk (i.e., probability) of committing a prediction error. Now, suppose you spend your night watching the game on TV, and that the first half is just over. The outcome of the first period may be treated as a random variable X, as well, possibly assuming any one of the same three values that Y may take, with

---

[1] US readers will forgive us for not picking up American football as an internationally popular sport. Nonetheless, they may appreciate the use of the word 'soccer' instead of 'football', tackling any legitimate misunderstandings.

A. Freno and E. Trentin: Hybrid Random Fields, ISRL 15, pp. 15–41.
springerlink.com   © Springer-Verlag Berlin Heidelberg 2011

probability distribution $P(X)$. Now, before the game starts over, the knowledge you acquired at break time may be used as 'observed evidence' that influences (i.e., refines) your prediction of the final value of Y. In fact, instead of relying on $P(Y)$ only, your prediction can now try to exploit the quantity $P(Y \mid X)$, namely the posterior probability of Y given the evidence X. If you repeat the calculation of $P(Y \mid X)$ in real-time while continuing watching the second half, i.e. as long as you acquire new evidence, the prediction becomes more and more likely. By the way, and technically speaking, after 45 minutes it is no longer a prediction, unless the referee goes for extra time. The question is: *how* can we use our prior knowledge, along with the new evidence, in order to come out with a mathematically sound prediction of the posterior probability? Answer to the question is yielded by the good, old Bayes' theorem[2] [73], which states that

$$P(Y \mid X) = \frac{P(X \mid Y)P(Y)}{P(X)} \qquad (2.1)$$

where the quantity $P(X \mid Y)$ is the likelihood of the evidence given Y, expressing how likely it is that the observation X complies with a specific outcome Y. Bayes' theorem is a sort of little magic that allows us to transform our prior probabilistic knowledge of an event into a more robust, posterior knowledge of its probability (given that new evidence has been observed, which has an influence on the likelihood of the event itself). As a major example, Bayes' theorem is the heart of statistical pattern recognition, where an object X is assigned to a class Y (out of a set of c possible, disjoint classes) based on the maximum a posteriori probability of Y given X (in fact, given a suitable representation of the very object).

The silly, yet effective, example of the soccer game implicitly conveys a generalization of the reasoning we built upon the bare application of Bayes' theorem. Indeed, a *train of consequences* could be hypothesized along the following line: we begin, before the match has started, with the prior probability of our team winning, losing, or drawing the match. Upon observation of evidence X_1 (e.g. the last minute breaking news that our goalie is sick and will be replaced by a 16-year old rookie) we calculate the new, posterior probability $P(Y \mid X_1)$ of the outcome given the evidence via Bayes' rule. Later on down the game new evidence X_2 is acquired, since it turns out that a penalty is assigned to the rival team. Let us assume, as in this instance, that X_1 and X_2 are independent of each other. Once again, we resort to Bayes' rule in order to further refine our prediction of the final outcome of the game given *all* the evidence collected so far, namely

$$\begin{aligned}
P(Y \mid X_1, X_2) &= \frac{P(X_1, X_2 \mid Y)P(Y)}{P(X_1, X_2)} \\
&= \frac{P(X_2 \mid Y)P(X_1 \mid Y)P(Y)}{P(X_2)P(X_1)} \\
&= \frac{P(X_2 \mid Y)P(Y \mid X_1)}{P(X_2)}
\end{aligned} \qquad (2.2)$$

---

[2] Please refer to Appendix A if you are unfamiliar with the fundamentals of Bayesian statistics.

where the independence between X_1 and X_2 has been exploited, the quantity $P(Y \mid X_1)$ was obtained above and it now plays the role of the 'prior', and the ratio $P(X_2 \mid Y)/P(X_2)$ expresses the impact that the new evidence X_2 has on the final outcome Y. The scheme can be iterated any time new evidence, say X_t, is observed at subsequent time t, allowing for a progressive refinement of the prediction. Let us define our prior confidence in a specific outcome, i.e. our subjective evaluation of probability $P(Y)$, as our *belief*. Then, the process we just described is an instance of what we can call a belief-propagation dynamics. Technically, it is referred to as Bayesian inference. Albeit rudimentary, the example above contains the germs of at least three major ideas: (*i*) there may be a causal relationship between random variables, which expresses an underlying semantic aspect (i.e., knowledge) of the phenomena under investigation; (*ii*) conditional probabilities of certain random variables given others (in particular, Bayes' posterior probabilities) are a suitable way of quantifying such causality relations in probabilistic terms; (*iii*) a belief (i.e., probability) can be 'propagated', leading to refined evaluations of posterior probabilities of certain variables given the observed value (or, the refined, propagated posterior probability) of others.

The very idea of a causal, binary relation between random variables implies, from an algebraic standpoint, the existence of a graph $\mathcal{G}$ that represents the relation itself. The vertexes of $\mathcal{G}$ are random variables, whilst edges represent conditional dependencies between pairs of such variables. A more intriguing (and, someways more realistic) example may help us developing this idea further. The example is a simplified version of a probabilistic expert system, suitable for assisting diagnostics or health-care applications. Consider the directed, acyclic graph shown in Figure 2.1.[3] Each vertex represents a particular condition of an individual. Conditions are thought of as discrete, binary variables having 'true' or 'false' as possible outcomes. For instance, 'Genetic predisposition to diabetes' means that the individual has/has not a family history of diabetes. This kind of genetic diathesis has a certain (prior) probability distribution. The meaning of the other vertices in the graph should be self-explanatory, accordingly. In real-world applications, priors for the different random variables may be estimated using frequentist, descriptive statistics over a representative sample of the population, accomplished via medical screening.

The edges in the graph, say (X,Y), have a fundamental, twofold meaning. In an abstract sense, they denote a causality relationship, meaning that the truth value of X exercises a certain causal effect on the likely outcome of Y. In a probabilistic sense, they imply statistical dependence between the variables X and Y, suitably expressed by means of (posterior) conditional probability distributions in the form $P(Y \mid X)$. For instance, the presence of kidney failure affects the likelihood of the patient undergoing a dialysis treatment, while this treatment is independent of the presence/absence of asthma. Obesity has not a direct causal effect on dialysis, but it increases the risk for diabetes mellitus which, in turn, renders renal failure more probable, eventually increasing the likelihood of a dialysis treatment altogether. More broadly, the probabilistic properties of a variable Y are affected by those of

---

[3] Please refer to Appendix B for a review of the fundamentals of graph theory.

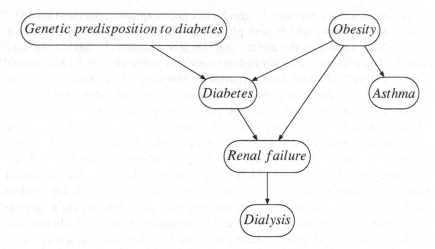

Fig. 2.1 An example of causal network concerning some medical conditions

the variables that are parents of X within the graphical structure. The very nature of this particular inferential mechanism suggests that we are undergoing an implicit Markov assumption.[4]

Application of this probabilistic graph to prediction or decision making requires a complete specification of two distinct elements: (*i*) the graphical structure, i.e. the random variables involved, and their mutual relationships of causality (that is, of statistical dependency). This confers a semantics to the model, and it requires prior knowledge of the domain provided by human experts in the field; (*ii*) the prior and the conditional probability distributions for all possible combinations of values of the random variables. This type of directed, acyclic graph is known as a *causal network*, which is also a first, particularly relevant instance of Bayesian network.

A sort of mystical aura seems to pervade the idea of Bayesian network, and of causal network in particular. Not only this is due to the fact that Thomas Bayes, who formulated a special case of the homonymic theorem in the eighteenth century [17], was a Presbyterian minister showing a deep, yet accidental interest in probability. In point of fact, no later than the year 2003 the physicist Stephen Unwin, in his book *The Probability of God* [307], uses Bayesian inference (and, implicitly, a causal network with a simple structure) in order to estimate the probability that God exists upon 'observations' of related evidence. Unwin estimates first the prior probability of God existing to be 0.5, as if flipping a coin. He then fixes (arbitrary) values for the likelihood of six different, major types of evidence given the existence of God. Instances of such evidence are: goodness of human beings, the evil that men do, intra- and extra-natural miracles, etc. Eventually, Bayesian inference is applied resulting in a 0.67 estimated probability which, via an additional bias rooted in faith,

---

[4] The Markov property, which is of the utmost importance to this book, will be formally introduced in Section 2.2.

adds up to 0.95. In our humble opinion, this underlines simply the pervasive effect of bias in statistics (providing us with further motivation in the development of the unbiased non-parametric estimation techniques that will be introduced in Chapter 5). Unwin's approach makes the adoption of the term 'belief propagation' particularly sound. It is noteworthy that another physicist, Larry Ford, applied the same causal network, evidence, and inference formulae, but using different estimates for the prior probabilities and the conditional likelihoods, coming up with a (possibly even sounder) estimate of the probability of God as large as 10^{-17} [290]. No doubt this instance of Bayesian statistics highlights the width of the range of its plausible applications. Moreover, it shows how the same graphical structure of a certain causal network may lead to completely different models and conclusions according to the prior and conditional probability distributions of individual variables.

The examples we made so far, and the very concept of causal network, may be somewhat misleading to the aim of understanding the general concept of Bayesian network. Indeed, they convey the ideas that (*i*) directed edges in the graph imply a causal relationship, and that (*ii*) all pairs of variables are dependent of each other if and only if a directed edge exists between them (in a way or the other). The general definition of Bayesian network, which will be given in a formal manner in Section 2.2, does not fulfill these expectations. It turns out that BNs are more general models, where the focus in on statistical *independencies* rather than *dependencies*, without any particular semantic assumption on possible causal relations among their variables. As a matter of fact, Bayesian networks are a simple, effective way to represent and calculate joint probabilities over a set of random variables under a Markov assumption. Exploiting the Markovian property (in order to obtain a viable factorization of the overall, joint distribution) requires the fundamental concept of d-separation, which relies on the idea of blocked paths within a graph (Section 2.2.1). Roughly speaking, the point of d-separation is that if all paths in the graph leading from one random variable (say, X) to another (say, Y) are blocked by a certain subset $\mathcal{S}$ of different variables, then X and Y are conditionally independent of each other given $\mathcal{S}$. The basic idea is refined, and finalized, by means of the notion of Markov blanket (as explained in Section 2.2.2), viz. a subset of the vertices of the graph that makes the distribution of a certain variable independent of all the others once its Markov blanket is given. This represents the building block of the desired factorization of the overall joint distribution.

Once the definition of Bayesian network is given, and the mechanisms of inference in Bayesian networks are pointed out, we are faced with a major problem: how can we fix the parameters that describe the conditional probability distributions associated with the random variables in the network? In the examples we made throughout this section we simply assumed that estimates of the probability distributions were (somewhat subjectively) known, expressing a sort of arbitrary belief. In real world scenarios, on the contrary, data are collected on the field, and the distributions are not known in advance. Section 2.3 covers this topic, taking a frequentist approach to parameter estimation from a given 'training' sample. Tricks for escaping the pitfalls of undersampled co-occurrences of specific outcomes of the variables are adopted, in order to skip singularities in the limit case, as well as to avoid numerical stability problems.

Parameter learning can come out with an accurate estimation of the conditional probability distributions, but it requires another form of prior knowledge, that is the graphical structure of the network. In other words, the graph which captures the (in)dependencies among the variables is expected to be fixed. Also this requirement is realistic only under specific circumstances, e.g. when the relationships among the variables do express some higher-level knowledge on the application domain, a knowledge which confers a meaningful semantics to the very network. Causal networks are relevant instances. In the examples above, the (causal) relation between pairs of variables was always clear, at least in an implicit manner. In the general case, as for parameter learning, application of Bayesian networks on the field cannot assume such a structural knowledge. For instance, in predicting the binding state of a cysteine within a protein it is assumed that the presence (or, absence) of a disulphide bond is affected by chemical/physical properties of other amino-acids at certain locations of the primary structure. Biologists do not know which amino-acids are actually involved, nor do they know what exact chemic-physical properties are relevant, or to which extent. A learning machine, e.g. a Bayesian network, is then expected to perform induction on a sample of proteins including examples of binding/non-binding cysteines, and to discover these relationships, in terms of statistical dependencies, on its own. Algorithms for learning the structure of Bayesian networks are presented in Section 2.4. Basically, they can be thought of as search procedures in the space of all possible graphs, where the search is guided by an optimality criterion expressed in terms of maximum likelihood of the network given the training sample (Section 2.4.1). The search for the optimal structure is accomplished according to a specific, statistically sound strategy (Section 2.4.2).

As we saw in Chapter 1, graphical models may become suitable tools for tackling pattern recognition problems. To this end, Section 2.5 introduces a simple, popular paradigm for the classification of patterns, namely the naive Bayes classifier, showing that it can be readily described (and mathematically justified) as a particular case of Bayesian network. This classifier will be used, among others, in order to obtain comparative experimental baselines in the applications presented later on in the book.

Finally, Section 2.6 summarizes the main topics of the chapter and draws conclusive remarks. In particular, it deals with extensions of the basic Bayesian network model, some of its applications, and offers a concise survey of the assets and limitations of Bayesian networks.

Aside from Thomas Bayes' early discoveries, the history of Bayesian networks is much more recent. Although some pioneering insights into the theory of BNs can be traced back to the work of Sewall Wright [318], the first systematic presentation of such models to the AI community is due to Judea Pearl [237]. While Pearl's work was concerned mainly with detailing the theoretical properties of BNs and developing efficient inference techniques, more recent work has been focusing on algorithms for learning Bayesian networks from data [50, 129, 286, 101].

2.2   Representation of Probabilities

Bayesian networks are used to represent joint probability distributions over sets of random variables. A Bayesian network is made up of two components: a directed acyclic graph (DAG), and a set of conditional probability tables (CPTs). Each node in the graph represents a random variable, and for each node there is a probability table specifying the conditional distribution of the variable given (each possible combination of) the values of its parents in the graph. A simple Bayesian network is exemplified in figure 2.2.

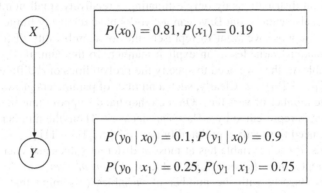

Fig. 2.2 A Bayesian network for the binary variables X and Y

In order to derive a joint probability distribution from a Bayesian network, the *directed Markov assumption* is made, according to which each variable is independent of its non-descendants in the DAG given the values of its parents. Consider the set $\mathbf{X}$ of random variables $X_1, \ldots, X_n$, and an arbitrary state $\mathbf{x} = x_1, \ldots, x_n$ of the variables in $\mathbf{X}$. If $\mathcal{PA}(X_i)$ is the set of parents of X_i, let $pa(X_i)$ denote the state of $\mathcal{PA}(X_i)$, i.e. some specific configuration of the values of the variables in $\mathcal{PA}(X_i)$. Then, the Markov assumption entails the following equality:

$$P(\mathbf{X} = \mathbf{x}) = \prod_{i=1}^{n} P(X_i = x_i | pa(X_i)) \qquad (2.3)$$

When X_i is a root node, $P(X_i = x_i \mid pa(X_i))$ refers to the absolute distribution of X_i, i.e. $P(X_i = x_i)$. Since the local distributions $P(X_i|pa(X_i))$ are provided by the conditional probability tables, Equation 2.3 specifies how to compute a joint probability distribution from a set of CPTs.

The derivation of Equation 2.3 from the Markov assumption can be carried out in the following way. Suppose that $X_1, \ldots, X_n$ are ordered *ancestrally*, where an ancestral ordering is any ordering such that, if node X_i is a parent of X_j, then $i < j$. Then, we can transform $P(x_1, \ldots, x_n)$ as follows:

$$P(x_1,\ldots,x_n) = \prod_{i=1}^{n} P(x_i \mid x_{i-1},\ldots,x_1)$$

$$= \prod_{i=1}^{n} P(x_i \mid pa(X_i)) \tag{2.4}$$

where the first step results from applying the chain rule,[5] and the second step exploits the Markov assumption in order to simplify each conditional probability $P(x_i \mid x_{i-1},\ldots,x_1)$ into $P(x_i \mid pa(X_i))$.

An advantage of using Bayesian networks is given by the fact that they allow to estimate joint distributions by only estimating a relatively small number of parameters. This also entails that Bayesian networks have relatively limited memory requirements. Suppose we want to represent the joint probability distribution of n (discrete) random variables in an explicit manner. To this aim, if $\mathcal{D}_i$ is the domain of variable X_i, then we need to specify the probabilities of d different events, where $d = (\prod_{i=1}^{n} \mid \mathcal{D}_i \mid) - 1$. Clearly, such a number of parameters grows exponentially with the number of variables. On the other hand, suppose that the probability distribution is represented by a Bayesian network. Then, the number of events for which we need to specify a probability is $d = \sum_{i=1}^{n} (\mid \mathcal{D}_i \mid - 1) \prod_{X_j \in \mathcal{PA}(X_i)} \mid \mathcal{D}_j \mid$. In other words, if each variable has at most m different values and each node has at most k parents, then the worst-case value of d is $n \cdot m^k \cdot (m - 1)$. In this case, d grows only linearly with the number of variables (assuming that k remains constant).

2.2.1 d-Separation

We now consider an important result in the theory of Bayesian networks, concerning the graphical property of *d-separation* [237, 226, 30]. This result provides a method for determining whether the graph of a Bayesian network entails any given conditional independence statement, i.e. a statement of the form 'X_i is independent of X_j given X_k'.

In order to define d-separation, we first need to introduce some terminology. If a DAG contains two edges (A,B) and (B,C), then we say that the edges (A,B) and (B,C) meet *head-to-tail* at node B, or that there is a *head-to-tail meeting* at node B between edges (A,B) and (B,C). If a DAG contains two arcs (A,B) and (A,C), then we say that the arcs (A,B) and (A,C) meet *tail-to-tail* at node A. If a DAG contains two arcs (A,B) and (C,B), then we say that the arcs (A,B) and (C,B) meet *head-to-head* at node B. We can now introduce the notion of *blocked chain*:[6]

Definition 2.1. Given a DAG $\mathcal{G} = (\mathcal{V}, \mathcal{E})$, if $A \in \mathcal{V}$, $B \in \mathcal{V}$, and $\mathcal{S} \subseteq \mathcal{V} \setminus \{A,B\}$, then a chain $\mathcal{C}$ connecting A and B is said to be *blocked* by $\mathcal{S}$ if $\mathcal{C}$ contains a node X such that one of the following conditions holds:

---

[5] See Theorem A.2 in Appendix A for a general statement and justification of the chain rule.
[6] See Appendix B.1 for a definition of chain.

1. $X \in S$ and the meeting at X in the considered chain is either head-to-tail or tail-to-tail;
2. $X \notin S$, S does not contain any descendant of X, and the meeting at X in the considered chain is head-to-head.

d-separation is then defined as follows:

Definition 2.2. Given a DAG $\mathcal{G} = (\mathcal{V}, \mathcal{E})$ such that $\{A, B\} \subset \mathcal{V}$ and $S \subseteq \mathcal{V} \setminus \{A, B\}$, the nodes A and B are said to be *d-separated* by S if all chains connecting A and B are blocked by S.

The importance of d-separation with respect to Bayesian networks is due to the following fact. If the DAG of a Bayesian network is such that two nodes A and B in the DAG are d-separated by a set S, then the Markov assumption entails that the variables A and B are independent given S [226]. That is, d-separation provides a general criterion for deriving conditional independence statements from Bayesian networks. A point that needs to be stressed is that d-separation is only a sufficient (and not a necessary) condition for conditional independence. Therefore, if A and B are not d-separated by S, we cannot conclude that A and B are not independent given S: it may be the case that a certain conditional independence holds in the distribution represented by a Bayesian network while that independence is not identified by d-separation. The assumption that d-separation is not only a sufficient, but also a necessary condition for conditional independence is usually referred to in the literature as *faithfulness assumption* [226]. In contrast to d-separation, the faithfulness property cannot be derived from the Markov condition, and therefore, while d-separation is a general property of Bayesian networks, faithfulness can only be assumed (if desired) as an additional condition. (The faithfulness assumption is made, for example, by constraint-based structure learning algorithms [286]).

The central role played by d-separation in the theory and application of Bayesian networks can be summarized by stating three theorems related to that property:

Theorem 2.1. *If a Bayesian network BN has DAG $\mathcal{G}$, then BN entails all and only those conditional independencies that are identified by d-separation in $\mathcal{G}$ (i.e. the set of conditional independence statements that can be derived from $\mathcal{G}$ through the d-separation criterion).*

Proof. See e.g. [226]. □

Before stating the second theorem, we introduce the concept of *Markov equivalence*:

Definition 2.3. Two Bayesian networks BN_1 and BN_2 with DAGs $\mathcal{G}_1 = (\mathcal{V}, \mathcal{E}_1)$ and $\mathcal{G}_2 = (\mathcal{V}, \mathcal{E}_2)$ are said to be *Markov equivalent* if they satisfy the following condition: for any pair of nodes A and B and any subset S of $\mathcal{V}$, A and B are d-separated by S in $\mathcal{G}_1$ if and only if A and B are d-separated by S in $\mathcal{G}_2$.

We can now state the second result:

Theorem 2.2. *Two Bayesian networks BN_1 and BN_2 with DAGs $\mathcal{G}_1$ and $\mathcal{G}_2$ entail the same set of conditional independencies if and only if they are Markov equivalent.*

Proof. The result is an immediate corollary of Theorem 2.1. $\square$

Finally, the following theorem (first proved in [238]) provides a practical way of recognizing whether two Bayesian networks are Markov equivalent:

Theorem 2.3. *Two Bayesian networks BN_1 and BN_2 with DAGs $\mathcal{G}_1$ and $\mathcal{G}_2$ are Markov-equivalent if and only if:*

1. *$\mathcal{G}_1$ and $\mathcal{G}_2$ contain the same* links, *i.e. the same arcs regardless of their direction;*
2. *$\mathcal{G}_1$ and $\mathcal{G}_2$ contain the same set of uncoupled head-to-head meetings.[7]*

Proof. See e.g. [226]. $\square$

2.2.2 Markov Blankets

First of all, let us define the general (statistical) concept of *Markov blanket* [237]:

Definition 2.4. Let $I(X,Y \mid Z)$ mean that X is independent of Y given Z, i.e. that $P(X \mid Y,Z) = P(X \mid Z)$. Then, if $\mathbf{X}$ is a set of random variables $X_1,\ldots,X_n$, a *Markov blanket* $\mathcal{MB}(X_i)$ for X_i in $\mathbf{X}$ is any subset $\mathcal{S}$ of $\mathbf{X} \setminus \{X_i\}$ such that $I(X_i,(\mathbf{X} \setminus \mathcal{S}) \setminus \{X_i\} \mid \mathcal{S})$.

In other words, the variables in $\mathcal{MB}(X_i)$ are such that $P(X_i \mid \mathbf{X} \setminus \{X_i\}) = P(X_i \mid \mathcal{MB}(X_i))$. A fundamental property of Bayesian networks is the one stated by the following theorem:

Theorem 2.4. *For any Bayesian network with nodes $X_1,\ldots,X_n$, if $\mathcal{PA}(X)$ denotes the set of parents of X and $\mathcal{CH}(X)$ denotes the set of children of X, then, for any node X_i, the following set is a Markov blanket of X_i in the network: $\mathcal{PA}(X_i) \cup \mathcal{CH}(X_i) \cup (\bigcup_{X_j \in \mathcal{CH}(X_i)} \mathcal{PA}(X_j) \setminus \{X_i\})$.*

Proof. See [237]. $\square$

That is to say, for each variable X_i, the set containing the parents, the children, and the parents of the children of X_i is sufficient in order to form a Markov blanket of X_i within the Bayesian network. An example is provided in Figure 2.3. Based on Theorem 2.4, every time we refer to *the* Markov blanket of a node X (as if it were warranted to be a unique set), we specifically mean the set $\mathcal{PA}(X) \cup \mathcal{CH}(X) \cup (\bigcup_{Y \in \mathcal{CH}(X)} \mathcal{PA}(Y) \setminus \{X\})$.

Being able to compute the distribution of a variable X_i given the state $mb(X_i)$ of its Markov blanket is very important when using Bayesian networks. For example, computing this distribution is a core component of the Markov chain Monte Carlo inference algorithm, which will be described in Section 4.4. More importantly, it

---

[7] The notion of uncoupled meeting is defined in Appendix B.1.

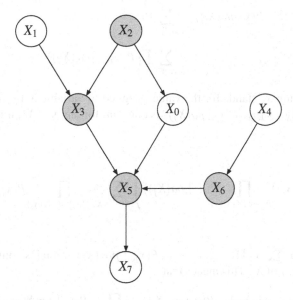

Fig. 2.3 The DAG of a Bayesian network representing the variables $X_0,\ldots,X_7$. In this network, a Markov blanket for X_0 is given by the set $\mathcal{MB}(X_0) = \{X_2,X_3,X_5,X_6\}$, denoted by shaded nodes.

will also play a key role in the use of the hybrid random field model (presented in Chapters 4–5). For any value x_i of X_i, this conditional distribution can be computed as follows [236]:

$$P(x_i \mid mb(X_i)) = \frac{P(x_i \mid pa(X_i))\prod_{X_j \in \mathcal{CH}(X_i)}P(x_j \mid pa(X_j))}{\sum_{x_{i_k} \in \mathcal{D}_{X_i}}P(x_{i_k} \mid pa(X_i))\prod_{X_j \in \mathcal{CH}(X_i)}P(x_j \mid pa(X_j),x_{i_k})} \quad (2.5)$$

where $P(x_j \mid pa(X_j),x_{i_k})$ is the same as $P(x_j \mid pa(X_j))$ except that we force X_i to assume value x_{i_k}. Equation 2.5 can be shown to hold true as follows. Let us write:

$$P(x_i \mid mb(X_i)) = \frac{P(x_i,mb(X_i))}{P(mb(X_i))} \quad (2.6)$$

Since the term $P(mb(X_i))$ does not depend on the particular value x_i of X_i, we can simplify equation 2.6 as follows:

$$P(x_i \mid mb(X_i)) \propto P(x_i,mb(X_i)) \quad (2.7)$$

Now, if $\mathcal{X}$ is the set of all possible joint states $\mathbf{x}$ of $X_1,\ldots,X_n$, let us denote by $\mathcal{X}^*$ the set of those states $\mathbf{x}$ such that $X_i = x_i$ and $\mathcal{MB}(X_i) = mb(X_i)$ in $\mathbf{x}$, where $\mathcal{X}^* \subset \mathcal{X}$. Clearly,

$$P(x_i, mb(X_i)) = \sum_{\mathbf{x} \in \mathcal{X}^*} P(\mathbf{x})$$

$$= \sum_{\mathbf{x} \in \mathcal{X}^*} \prod_{j=1}^{n} P(x_{j_\mathbf{x}} \mid pa_\mathbf{x}(X_j)) \qquad (2.8)$$

where the notation $x_\mathbf{x}$ stands for the value assigned to variable X by state $\mathbf{x}$. Since $P(x_i \mid pa(X_i)) \prod_{X_j \in \mathcal{CH}(X_i)} P(x_j \mid pa(X_j))$ is constant for any $\mathbf{x} \in \mathcal{X}^*$, it follows that

$$P(x_i, mb(X_i)) =$$

$$= \left(P(x_i \mid pa(X_i)) \prod_{X_j \in \mathcal{CH}(X_i)} P(x_j \mid pa(X_j)) \right) \sum_{\mathbf{x} \in \mathcal{X}^*} \prod_{X_j \notin \{X_i\} \cup \mathcal{CH}(X_i)} P(x_{j_\mathbf{x}} \mid pa_\mathbf{x}(X_j))$$

$$(2.9)$$

Clearly, the term $\sum_{\mathbf{x} \in \mathcal{X}^*} \prod_{X_j \notin \{X_i\} \cup \mathcal{CH}(X_i)} P(x_{j_\mathbf{x}} \mid pa_\mathbf{x}(X_j))$ remains constant for any particular value x_i of X_i. This means that

$$P(x_i, mb(X_i)) \propto P(x_i \mid pa(X_i)) \prod_{X_j \in \mathcal{CH}(X_i)} P(x_j \mid pa(X_j)) \qquad (2.10)$$

which is exactly what we need in order to justify equation 2.5.

2.3 Parameter Learning

In Bayesian networks, parameter learning is the problem of learning the conditional probability tables from data, assuming a fixed network structure (i.e. assuming the DAG is known). To this aim, we compute a set of relative frequencies and then we use these relative frequencies as estimates of the relevant probabilities. The set of relevant probabilities/relative frequencies is clearly determined by the network structure: for each variable, we need to estimate its conditional distribution for each particular state of its parents in the DAG. (For the DAG roots, we simply estimate their unconditional probabilities).

More formally, suppose we have a dataset $\mathbf{D}$ containing m data points, where each data point $\mathbf{d}_j$ is a vector $(x_{1_j}, \dots, x_{n_j})$ of values of the variables $X_1, \dots, X_n$. The simpler case in parameter learning is the case of a variable X_i having no parents in the DAG. In this case, we only need to estimate the (unconditional) distribution $P(X_i)$. For each value x_{i_k} of X_i, our estimate will be the following:

$$\widehat{P}(X_i = x_{i_k}) = \frac{|\{\mathbf{d}_j : X_i = x_{i_k}\}|}{m} \qquad (2.11)$$

where the notation $\widehat{P}(X = x)$ refers to our estimate of $P(X = x)$.

The more general case is the case of learning the conditional distribution of a node X_i having parents $\mathcal{PA}(X_i)$. In this case, we need to estimate a distribution $P(X_i \mid pa(X_i))$ for each possible state $pa(X_i)$ of $\mathcal{PA}(X_i)$. For each value x_{i_k} of X_i, we will estimate the conditional probabilities as follows:

$$\widehat{P}(X_i = x_{i_k} \mid pa(X_i)) = \frac{|\{\mathbf{d}_j : X_i = x_{i_k} \wedge \mathcal{PA}(X_i) = pa(X_i)\}|}{|\{\mathbf{d}_j : \mathcal{PA}(X_i) = pa(X_i)\}|} \tag{2.12}$$

The strategy proposed in equations 2.11–2.12 suffers from the following problem. If a particular value x_{i_k} of X_i is never observed in $\mathbf{D}$, or if it is never observed together with a particular configuration $pa(X_i)$ of $\mathcal{PA}(X_i)$, then our estimate of $P(X_i = x_{i_k})$ (or of $P(X_i = x_{i_k} \mid pa(X_i))$) will be zero. This result is not acceptable in all cases where any event is possible, i.e. where every state of the network can be observed in principle. A solution for this difficulty appeals to the notion of an *equivalent sample size* [217], which we denote by N. The equivalent sample size is the size of a theoretical sample (the 'equivalent sample') which we assume to have been observed before the actual dataset $\mathbf{D}$. In other words, it is the size of a *prior* sample. Within this prior sample, we assume to have observed any particular value x_{i_k} of X_i for a number of times equal to $p \cdot N$, where p stands for the prior probability that X_i has value x_{i_k}.

Going back to equations 2.11–2.12, we revise them as follows:

$$\widehat{P}(X_i = x_{i_k}) = \frac{|\{\mathbf{d}_j : X_i = x_{i_k}\}| + p_{i_k}N}{m + N} \tag{2.13}$$

$$\widehat{P}(X_i = x_{i_k} \mid pa(X_i)) = \frac{|\{\mathbf{d}_j : X_i = x_{i_k} \wedge \mathcal{PA}(X_i) = pa(X_i)\}| + p_{i_k}N_{pa_i}}{|\{\mathbf{d}_j : \mathcal{PA}(X_i) = pa(X_i)\}| + N_{pa_i}} \tag{2.14}$$

where N_{pa_i} is a parameter related to N in a way we will explain shortly. An important question concerning equations 2.13–2.14 is what values we choose for the parameters p_{i_k}, N, and N_{pa_i}. In typical applications (and throughout the applications of Bayesian networks presented in Chapter 6), we assign uniform prior probabilities to the different values of each variable. Therefore, our choice for p_{i_k} will be the following:

$$p_{i_k} = \frac{1}{|\mathcal{D}_i|} \tag{2.15}$$

where $\mathcal{D}_i$ is the domain of variable X_i. The value we assign to N is instead:

$$N = \max_{1 \leq i \leq n} |\mathcal{D}_i| \tag{2.16}$$

An intuitive justification of Equation 2.16 appeals to two different aims: on the one hand, we want to keep the equivalent sample size as small as possible, so as to prevent prior probabilities from biasing learning too heavily; on the other hand, we want the equivalent sample size to be large enough to contain at least one occurrence for all values of each variable. Therefore, the choice made in equation 2.16 seems to be a reasonable compromise. Given N, we define N_{pa_i} as follows:

$$N_{pa_i} = \frac{N}{|\mathcal{D}_{\mathcal{PA}_i}|} \qquad (2.17)$$

where $\mathcal{D}_{\mathcal{PA}_i}$ is the set of all possible states of $\mathcal{PA}(X_i)$. As a result of Equation 2.17, each value x_{i_k} of a non-root node X_i turns out to have been observed exactly $\frac{N}{|\mathcal{D}_i|}$ times within the equivalent sample, where $\frac{N}{|\mathcal{D}_i|} \geq 1$ [226]. To realize why $\frac{N}{|\mathcal{D}_i|} \geq 1$, we have to keep in mind that x_{i_k} is observed $p_{i_k} N_{pa_i}$ times for each possible state $pa(X_i)$ of $\mathcal{PA}(X_i)$. In other words, we have that $\frac{N}{|\mathcal{D}_i|} = p_{i_k} N_{pa_i} \cdot |\mathcal{D}_{\mathcal{PA}_i}|$.

2.4 Structure Learning

While parameter learning requires that the DAG of the Bayesian network has been previously specified, structure learning aims at inferring from a dataset the DAG itself (together with the conditional probability tables). A general way of formalizing this task consists in viewing it as a *search* problem. In this setting, if we are dealing with the random variables $X_1, \ldots, X_n$, then the *problem space* is given by the set of all possible DAGs with nodes $X_1, \ldots, X_n$, and the task is to find the DAG such that the corresponding Bayesian network maximizes a given *evaluation function*. Therefore, given the search space, what we have to specify is first a suitable evaluation function, and second a *search strategy* allowing us to efficiently explore that space.

2.4.1 Evaluation Function

As we saw in section 2.2, given a Bayesian network h, equation 2.3 specifies how to compute the probability of a pattern $(x_1, \ldots, x_n)$. In other words, equation 2.3 implicitly specifies the value of the conditional probability $P(x_1, \ldots, x_n | h)$. If we are given not just one pattern $\mathbf{x}$, but a set $\mathbf{D}$ of patterns $\{\mathbf{x}_1, \ldots, \mathbf{x}_m\}$, then, by assuming that the patterns in $\mathbf{D}$ are independent given the underlying distribution, we can write down the joint probability of the dataset given a Bayesian network h in the following way:

$$
\begin{aligned}
P(\mathbf{D} \mid h) &= \prod_{j=1}^{m} P(\mathbf{x}_j \mid h) \\
&= \prod_{j=1}^{m} \prod_{i=1}^{n} P(X_i = x_{i_j} \mid pa_j(X_i))
\end{aligned}
\qquad (2.18)
$$

where $pa_j(X_i)$ stands for the state of the parents of X_i as it is determined by pattern $\mathbf{x}_j$. It is important to realize that the probability $P(X_i = x_{i_j} \mid pa_j(X_i))$ is conditional with respect to h, since the parents of each node X_i cannot be identified without assuming a particular network structure.

A reasonable choice for the evaluation function is a Bayesian scoring function [226], aimed at measuring the posterior probability of the Bayesian network given the data. Since we know how to compute $P(\mathbf{D} \mid h)$, i.e. the likelihood of a Bayesian network h given the data, the Bayesian score can be computed as follows:

$$P(h \mid \mathbf{D}) = \frac{P(\mathbf{D} \mid h)P(h)}{P(\mathbf{D})} \tag{2.19}$$

where $P(h)$ is the prior probability of h. This transformation derives from applying Bayes' theorem. Since, for a fixed dataset $\mathbf{D}$, $P(\mathbf{D})$ remains constant for any h, we can drop this term from equation 2.19 for the purposes of ranking different networks by means of their Bayesian score. Thereby we obtain the following equation:

$$P(h \mid \mathbf{D}) \propto P(\mathbf{D} \mid h) \cdot P(h) \tag{2.20}$$

The main difficulty with the Bayesian scoring function lies in estimating the prior probability expressed by $P(h)$. A typical scenario is the one where the priors of all different networks (for a given set of variables) are unknown, and they are taken to be uniform. If we assume uniform prior probabilities, then the Bayesian scoring function given in equation 2.20 collapses into the likelihood function $P(\mathbf{D} \mid h)$. In this case, our evaluation function will lead to maximum likelihood hypotheses.

A pure maximum likelihood strategy is not suitable for structure learning, for two different reasons. First, a practical reason is that maximum likelihood structure learning leads to overfitting the training data. Overfitting in learning the structure of Bayesian networks typically takes the form of a bias towards *complete* DAGs (i.e. DAGs which are fully connected). In fact, the more edges there are in the DAG, the more parameters there will be in the resulting Bayesian network, which means that the Bayesian network will be more *expressive*. This higher representational power leads the learning process to adapt the DAG to the noise contained in the data, since capturing that noise will increase the probability of the data given the network (i.e. the network likelihood). But networks that overfit the training data will generalize poorly to test data. Evaluation of the likelihood on a separate validation set would be required, at the very least.

The second reason against the maximum likelihood scoring function has a more formal nature [226]. In order to understand the argument, we first need to recall that different DAGs can be equivalent with respect to the Markov condition (as explained in Section 2.2.1). For example, the DAG in Figure 2.2 is Markov-equivalent to a DAG (with nodes X and Y) where the edge from X to Y is reversed (i.e. where Y is parent of X), because both DAGs entail no independencies based on the Markov assumption. Actually, this equivalence also follows from the simple fact that $P(X)P(Y \mid X) = P(X,Y) = P(Y)P(X \mid Y)$, i.e. that the joint distribution of X and Y is factorized equivalently in the two networks. Given this, it follows that when we search a space of candidate DAGs, we are actually searching a space of *equivalence classes* containing Markov-equivalent DAGs. Therefore, if we assign uniform priors to all possible DAGs (as the maximum likelihood strategy entails), we end up assigning higher priors to larger equivalence classes. For example, given n nodes, there is only one DAG entailing the complete set of independencies (i.e. the empty DAG), while there are $n!$ equivalent DAGs entailing no independencies (i.e. all complete DAGs). This means that a uniform priors assumption would end up assigning to the latter kind of distribution a prior probability which is $n!$ times higher than the prior probability assigned to the former kind of distribution. Actually, a preliminary step for a correct implementation of maximum likelihood structure learning in BNs should consist in assigning uniform prior probabilities to Markov-equivalence classes of DAGs, rather than to the DAGs themselves. Clearly, such a preprocessing step would be computationally intractable, due to the difficulty of identifying all the involved Markov-equivalence classes (so as to compute the prior of each equivalence class).

The reasons given above enforce the worry that the search for the best scoring Bayesian network may be heavily biased toward densely connected graphs. One way to overcome the difficulties of (the maximum likelihood version of) the Bayesian scoring function is offered by the *minimum description length* principle ([259]). According to this principle, we seek the Bayesian network that maximizes the likelihood while minimizing the length of the Bayesian network description. The length of describing a Bayesian network is nothing but the length of its encoding using a specified language. The version of the MDL principle that we will use for structure learning takes the form of the heuristic function $MDL(h)$, defined as follows [317]:

$$MDL(h) = \log P(\mathbf{D} \mid h) - \frac{DL(h)}{2} \log |\mathbf{D}| \qquad (2.21)$$

where $DL(h)$ is the number of parameters specified in the Bayesian network h. $MDL(h)$ penalizes the likelihood of h to an extent that is proportional to the network complexity, where complexity is measured by $DL(h)$. The aim of this heuristic is to encourage introducing parameters when the parameters really capture regularities in the data (and hence produce a strong increase of the network likelihood), and to discourage parameter introduction when this only captures the noise in the data (and hence increases the likelihood to a relatively small extent). In other words, the idea is to maximize the likelihood while keeping the DAG as sparse as possible. The MDL evaluation function is also referred to as *Bayesian information criterion* approximation [271]. It can be shown that the MDL/BIC scoring function is asymptotically correct [271, 226].

It is interesting to note that the score assigned to h by the MDL/BIC score is tightly related to the posterior probability of h. This remark derives from the following argument [217]:

$$
\begin{aligned}
\arg\max_{h} P(h \mid \mathbf{D}) &= \arg\max_{h} \frac{P(\mathbf{D} \mid h)P(h)}{P(\mathbf{D})} \\
&= \arg\max_{h} P(\mathbf{D} \mid h)P(h) \\
&= \arg\max_{h} \{\log_2 P(\mathbf{D} \mid h) + \log_2 P(h)\} \\
&= \arg\min_{h} \{-\log_2 P(\mathbf{D} \mid h) - \log_2 P(h)\}
\end{aligned}
\tag{2.22}
$$

As originally shown in [276], the result of derivation 2.22 can be interpreted as stating that a way of maximizing the posterior probability of h is by minimizing the sum of the length (in bits) of encoding the data given the information provided by h and the length (in bits) of encoding h. While the first quantity corresponds to the likelihood of h, the second quantity corresponds to the prior probability of h, which means that decreasing the length of the model description increases the prior probability of the model. Clearly, the number of parameters specified in a network is a measure of the network description length: therefore, the prior probability of a network is inversely proportional to the number of its parameters. Of course, when we are not able to assess $P(h)$, we cannot be able to assess $\log_2 P(h)$ either. For this reason, we use a heuristic based on the MDL principle, since we cannot measure in the strict sense the description length of models. In other words, the MDL heuristic offers a well-grounded way to approximate the prior probability of different candidate models when no exact estimate of priors is available.

2.4.2 Search Strategy

We now present two very popular search strategies for learning the structure of Bayesian networks. Section 2.4.2.1 describes a general hill-climbing approach, while the so-called K2 algorithm is described in Section 2.4.2.2.

2.4.2.1 Hill-Climbing Search

Once a suitable evaluation function has been defined, we can specify a strategy for searching the model space. In order to find the model with the highest MDL score, a relatively efficient method consists in using a hill-climbing algorithm [317]. The algorithm starts from a given Bayesian network h (typically, from a BN with a DAG containing no edges), and it generates a set of 'neighbors' of h, where a neighbor is a Bayesian network whose DAG is obtained by operating on h in one of three ways: (*i*) adding one arc, (*ii*) removing one arc, or (*iii*) reversing one arc. Once a new DAG has been constructed, a corresponding Bayesian network is obtained by learning the conditional probability tables for that DAG. While generating new DAGs, we also need to take care to discard the ones containing cycles. The neighbors of h are scored, and the highest-scoring neighbor h^* is compared to h. If h^* scores better than

h, the whole cycle is iterated for h^*, and the process continues until no neighbor of the currently accepted network improves on the score of that network. Since the search algorithm is greedy, it may be run several times, starting each time from a different, randomly generated DAG, so as to reduce the risk of getting stuck into local maxima. In order to speed up the search, a useful trick is to maintain a *tabu list* [114], keeping track of the states explored at each iteration, so as to prevent the algorithm from scoring several times models already encountered during the search. The pseudocode of Algorithm 2.1 implements the strategy just described.

Algorithm 2.1 HCLearnBN : Hill-climbing structure learning in Bayesian networks

Input: Dataset **D**; initial Bayesian network h.
Output: Bayesian network h maximizing the heuristic function $MDL(h)$ with respect to **D**.

HCLearnBN(**D**, h):
1.   do
2. $s = MDL(h)$
3. $\mathcal{H} = \{h_i : h_i$ is a neighbor of $h\}$
4. $h^* = \arg\max_{h_i \in \mathcal{H}} MDL(h_i)$
5. $s^* = MDL(h^*)$
6.   if $(s^* > s)$
7. $h = h^*$
8.   while $(s^* > s)$
9.   return h

2.4.2.2   The K2 Algorithm

Another greedy search strategy (which is faster than Algorithm 2.1) is proposed in [50], where the so-called K2 algorithm is developed by evolving a previous method named *Kutató* [132]. The K2 training algorithm must be provided with an ordering over the network nodes (which will be ancestral with respect to the learned DAG) and with a parameter k, indicating the maximum number of parents allowed for each node. These two inputs can be either fixed on the basis of prior knowledge (if available) or tuned by means of cross-validation. Starting from an empty DAG, each node is assigned as parents the k nodes such that the resulting DAG maximizes a specified scoring function, where the parents of each node X_i are selected from the set $\mathcal{X}_i = \{X_j : j < i\}$. However, arcs are not added whenever augmenting the DAG has the effect of decreasing the network score. This means that, in the learned network, each node will have at most k parents (but possibly fewer). Pseudocode for the K2 structure learning method is given by Algorithm 2.2, where the log-likelihood function is used as scoring metric.

In principle, any suitable evaluation function may be chosen according to the aims of each application. However, when the ordering over the nodes and the value of the k parameter are set in a suitable way, the K2 search technique is less prone to overfitting than the general hill-climbing algorithm described earlier, since it usually

---

Algorithm 2.2 K2LearnBN: K2 structure learning in Bayesian networks

---

Input: Dataset $\mathbf{D}$; Bayesian network h with DAG $\mathcal{G} = (\mathcal{V}, \mathcal{E})$, where $\mathcal{V} = \{X_1, \ldots, X_n\}$ and $\mathcal{E} = \emptyset$; integer k.

Output: Bayesian network h maximizing the log-likelihood $\log P(\mathbf{D} \mid h)$.

K2LearnBN$(\mathbf{D}, h, k)$:
```
 1.    for(i = 1 to n)
 2.        𝒫𝒜(Xᵢ) = ∅
 3.        addMoreParents = true
 4.        while(addMoreParents ∧ |𝒫𝒜(Xᵢ)| < k)
 5.            𝒳 = {Xⱼ : j ≤ i}
 6.            Xⱼ = argmax
 7.            if
 8.                𝒫𝒜(Xᵢ) = 𝒫𝒜(Xᵢ) ∪ {Xⱼ}
 9.            else
10.                addMoreParents = false
11.        for(Xⱼ ∈ 𝒫𝒜(Xᵢ))
12.            𝒠 = 𝒠 ∪ {(Xⱼ, Xᵢ)}
13.    return h
```

5. $\quad \mathcal{X} = \{X_j : j \le i\}$

6. $\quad X_j = \arg\max_{X_j \in \mathcal{X} \setminus \mathcal{P}\mathcal{A}(X_i)} \log \sum_{l=1}^{|\mathbf{D}|} P\big(x_{i_l} \mid pa_l(X_i), x_{j_l}\big)$

7. $\quad \mathbf{if}\,(\log \sum_{l=1}^{|\mathbf{D}|} P\big(x_{i_l} \mid pa_l(X_i), x_{j_l}\big) > \log \sum_{l=1}^{|\mathbf{D}|} P\big(x_{i_l} \mid pa_l(X_i)\big))$

8. $\quad\quad \mathcal{P}\mathcal{A}(X_i) = \mathcal{P}\mathcal{A}(X_i) \cup \{X_j\}$

---

prevents the DAG from being augmented beyond necessity. On the other hand, the K2 algorithm is likely to deliver less accurate results than Algorithm 2.1 when no prior knowledge is available concerning the application domain, since in this case it is not at all trivial to determine optimal choices both for the topological ordering of the nodes and the value of the k parameter. Although the K2 algorithm is usually much faster than Algorithm 2.1 (at the cost of being less accurate), it may still be very expensive to run for high-dimensional problems. In fact, it is shown in [50] that its worst-case computational complexity is $O(n^4)$.

2.5 The Naive Bayes Classifier

An interesting kind of directed graphical model is the so-called *naive Bayes classifier* [70]. The naive Bayes (NB) model is a particular case of Bayesian network, especially designed for application to pattern classification. If we deal with a collection of patterns described by features $X_1, \ldots, X_d$ and partitioned into classes $c_1, \ldots, c_n$, then a naive Bayes classifier for the considered domain is a Bayesian network with DAG $\mathcal{G} = (\mathcal{V}, \mathcal{E})$, where $\mathcal{V} = \{X_1, \ldots, X_d, C\}$, C is a random variable ranging over the set $\{c_1, \ldots, c_n\}$ (hence denoting the class of a given pattern), and $\mathcal{E} = \{(C, X_i) : 1 \le i \le d\}$. In other words, the naive Bayes model for $\mathcal{V}$ assumes that, for any pair of features X_i and X_j, $P(X_i \mid X_j, C) = P(X_i \mid C)$, i.e. X_i is independent of X_j once the value of C is known. As an example, the graph of a naive Bayes classifier is illustrated in Figure 2.4, where $d = 4$.

In order to explain how patterns are classified according to the naive Bayes technique, we first need to recall a basic principle of the Bayesian decision theory. Given

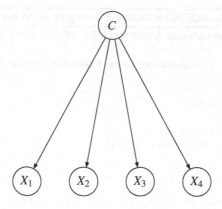

Fig. 2.4 The graphical structure of a naive Bayes model for a classification problem with features $X_1, \ldots, X_4$ and class C

a pattern $\mathbf{x} = (x_1, \ldots, x_d)$, the class having the *maximum a posteriori* (MAP) probability of containing $\mathbf{x}$ is given by

$$c^{MAP} = \underset{c_i \in \mathcal{D}_C}{\arg\max} \, P(c_i \mid x_1, \ldots, x_d) \qquad (2.23)$$

where $\mathcal{D}_C$ is the domain of variable C. If we adhere to the principles of Bayesian classification, we should endorse c^{MAP} as our decision for any new pattern to be classified. Using Bayes theorem, Equation 2.23 can be rewritten as follows:

$$
\begin{aligned}
c^{MAP} &= \underset{c_i \in \mathcal{D}_C}{\arg\max} \, \frac{P(x_1, \ldots, x_d \mid c_i) P(c_i)}{P(x_1, \ldots, x_d)} \\
&= \underset{c_i \in \mathcal{D}_C}{\arg\max} \, P(x_1, \ldots, x_d \mid c_i) P(c_i)
\end{aligned}
\qquad (2.24)
$$

Given the structure of the naive Bayes model, it is clear that any ordering over the features $X_1, \ldots, X_d$ is ancestral. Therefore, if we apply first the chain rule and then the Markov assumption (in the same way we applied them in Equation 2.4), we come up with the following result:

$$
\begin{aligned}
c^{MAP} &= \underset{c_i \in \mathcal{D}_C}{\arg\max} \, P(c_i) \prod_{j=1}^{d} P(x_j \mid x_{j-1}, \ldots, x_1, c_i) \\
&= \underset{c_i \in \mathcal{D}_C}{\arg\max} \, P(c_i) \prod_{j=1}^{d} P(x_j \mid c_i)
\end{aligned}
\qquad (2.25)
$$

Based on Equation 2.25, in order to train the naive Bayes model for a specified classification task, we simply need to estimate the parameters of each conditional distribution $P(X_i \mid C)$, for $1 \leq i \leq d$. To this aim, we use the technique described in Section 2.3.

If we compare Equation 2.25 to Equation 2.24, we immediately understand the main advantage of using the naive Bayes classifier. While computing the conditional distribution $P(X_1, \ldots, X_d \mid C)$ requires us to estimate $O(m^d - 1)$ parameters, where $m = \max_{1 \leq i \leq d} |\mathcal{D}_{X_i}|$, computing instead the product $\prod_{j=1}^{d} P(x_j \mid c_i)$ requires us to estimate only $O(m - 1)$ parameters. Therefore, parameter estimation in a naive Bayes model will be much more robust as compared to a direct estimate of the conditional distribution specified in Equation 2.24.

Surprisingly, although the independence assumption involved in the naive Bayes model may seem to be too strong at first glance, it is shown in [70] that the described classification technique can be optimal (with respect to misclassification rate) even when the independence assumption is violated to a significant extent. This result explains why the naive Bayes classifier performs unexpectedly well in several applications where the features are known to display some relevant pattern of dependencies. Another (theoretical and experimental) analysis of the optimality properties of naive Bayes classification is performed in [126]. Several efforts have also been devoted by the research community to the task of weakening the independence assumption at issue, without losing the efficiency and robustness properties of the standard naive Bayes method. Some of these techniques are investigated in [254, 312].

2.6 Final Remarks

In summary, this chapter reviewed Bayesian networks and their fundamental mathematical and algorithmic background. We saw that Bayesian networks are a particularly relevant instance of the generic graphical model, a perfect starting point for investigating the idea of using a relational structure between pairs of random variables which can be represented in the form of a graph, with the understanding that a directed edge between two variables may exist only if the variables are not statistically independent of each other. On the other way around, correlation between the variables does not entail the presence of an edge between them. In fact, the (mutual) statistical dependence may be expressed in terms of the influence that other, intermediate variables can exercise or undergo. This generic concept, enforced by the Markov assumption, is formally encapsulated within the notions of d-separation and, in turn, of Markov blanket. The definitions and properties of these will be of the utmost relevance in the very idea of hybrid random field, that shall be introduced in Chapter 4.

We also saw that the probabilistic characterization of the graphical structure in Bayesian networks is completed by an explicit expression of the conditional probability distributions associated with each variable in the graph, by means of conditional probability tables. The CPT for a generic variable X lists the probability of all possible values that X can take, given any values the parents of X (according to the graphical structure) can take. In the general case, these probabilities are not known in advance. They shall rather be estimated from a training sample of data collected on the field. In this respect, a Bayesian network having a given, fixed structure can be seen as a particular parametric model, whose parameters are the entries of the CPTs. Maximum-likelihood parameter estimation, relying on the frequentist approach of counting co-occurrences of the outcomes of the variables over the training sample, is a popular, effective technique for parameter learning in Bayesian networks.

In several scenarios the graphical structure of a Bayesian network is expected to be fixed. In these cases, prior knowledge by a human expert is needed in order to fix the directed acyclic graph. This is typical of probabilistic expert systems, for instance. Under such circumstances, the graph expresses some underlying semantics on the application domain. Causal networks are a particular type of Bayesian networks that fulfills this requirement, where the presence of a directed edge in the graph states the existence of a causal relationship between two variables. In general, unfortunately, this kind of prior, structural knowledge is unavailable. For this reason, algorithms for learning the graphical structure in a Bayesian network are needed. We saw how structure learning from a data sample can be readily defined as a search problem in the space of possible graphs over a given set of variables. The search strategy relies on a probabilistic criterion function, namely the likelihood of the model given the data.

Although this chapter introduced Bayesian networks defined over traditional, discrete random variables, extensions to continuous-valued variables can be found in the literature [136, 189, 52]. We will investigate the issue of real-valued probability distributions in depth in Chapter 5, where hybrid random fields for continuous spaces are introduced.

2.6.1 Extensions of Bayesian Networks

Several extensions to the basic Bayesian network model have been proposed in the literature. Object oriented Bayesian networks (OOBN) are basically stochastic, functional object oriented languages [175]. In OOBNs classes and objects are represented in terms of BNs, where each vertex in the graph is an attribute of the object itself (which, in turn, can be treated as another object) and the edges represent probabilistic relationships between objects/attributes. An inference algorithm for OOBNs is proposed in [175], whilst parameter learning for OOBNs is discussed in [188], as well.

Another extension to Bayesian networks can be pursued in terms of *abstraction* and *aggregation* [213, 214]. Abstraction (also known as 'state space abstraction') is the operation of replacing diverse, possible states of a certain vertex with a single, unifying meta-state for the node itself [313, 46]. Aggregation ('structural abstraction'), in a conceptually similar but technically different framework, replaces several vertexes in the BN graph with a unique meta-node [313, 287]. Refinement and decomposition are the natural, opposite operations, i.e. they move one step down the hierarchy of meta-levels of abstraction. Applications of these operations may ease the task of representation, interpretation and acquisition of knowledge in Bayesian networks. Moreover, they may lead to improved performance (in terms of computational speed) of the inference mechanisms [213, 214].

In [89] Bayesian networks are extended in order to suitably represent regression models. As we saw, a Bayesian network is intrinsically a model of a joint distribution over a set of random variables. In regression problems, we are interested in modeling the conditional distribution of a specific variable (or, of a subset of the variables) given the others. In so doing, we are allowed to make predictions on the expected value of certain random quantities as a function of the observed outcomes of others. For instance, a biologist would like to predict the expected period of time a given molecule (having specific chemical and physical characteristics) is going to take before bio-degrading in a specific environment (e.g., in the presence of water). The approaches presented in [89] are viable BN-based solutions to the regression problem, allowing also for the combination of discrete- and continuous-valued random variables.

Major extensions to Bayesian networks concern the representation and processing of time sequences. Roughly speaking, in sequence modeling each random variable is indexed by a discrete time index, e.g. we write X_t instead of the bare X. The time index t is expected to range over the interval $0 \leq t \leq T$, where $T + 1$ is the length of the sequence under consideration. According to the nature of the application, T may be constant (i.e., all sequences in the problem domain share the same length) or, more realistically, T is variable and sequence-specific. This general case is met in popular real-world scenarios, ranging from speech processing (where a sequence of random vectors of acoustic features is extracted at fixed time intervals, and it represents the characteristics of the speech signal uttered by a speaker as it evolves over time) to handwriting recognition, video processing, and bioinformatics (where the index t is a positional index which runs over an input sequence of individual bio-chemical items, e.g. the amino-acids which build up the primary structure of a protein). Dynamic Bayesian networks (DBN) are a major instance of Bayesian networks extended to sequences [110]. From a general standpoint, DBNs are simply BNs where time-indexed variables $X_0, \ldots, X_T$ are treated (and, represented within the graphical structure) as separate variables (i.e., as individual vertexes in the graph), and the edges—being of causal nature—follow the natural time order (e.g., a directed edge exists between X_t and X_{t+1}). The model may also take into account *hidden* state variables, referring to an underlying, non-observable random process which evolves over time and that is responsible for the dynamics of the observable state variables (either discrete, or continuous). Although DBNs are a very general, unifying framework for the probabilistic representation of time series, some problems arise at the algorithmic level when it comes down to inference and learning. In fact, probabilistic inference and, in turn, parameter learning, turn out to be intractable in DBNs [110]. Practical work-arounds are outlined in [110] in terms of tractable variational approximations aimed at the maximization of a lower-bound of the overall likelihood criterion. It is a fact of the utmost interest that several, popular paradigms can be interpreted as particular (yet, simple) instances of DBNs, namely Kalman filters [158, 264], hidden Markov models [251], and Input-output hidden Markov models [21]. Due to their theoretical and practical relevance, hidden Markov models (HMMs) are worth reviewing here, pinpointing the way they can be seen as BNs.

A hidden Markov model is a pair of stochastic processes: a *hidden* Markov chain and an *observable* process which is a probabilistic function of the states of the former. This means that observable events in the real world (e.g., the amino-acids along a protein primary structure) are modeled with (possibly continuous) probability distributions, that are the observable part of the model, associated with individual states of a discrete-time, first-order Markov process. The semantics of the model (conceptual correspondence with physical phenomena) is usually encapsulated in the hidden part. For instance, in automatic speech recognition a hidden Markov model can be used to model a word in the task-dependent vocabulary, where each state of the hidden part represents a phoneme (or sub-phonetical unit), whereas the observable part accounts for the statistical characteristics of the corresponding acoustic events in a given feature space (e.g. sampled acoustic signal, represented in a proper way). More precisely, a hidden Markov model is defined by:

1. A set S of Q states, $S = \{S_1, \ldots, S_Q\}$, which are the distinct values that the discrete, hidden stochastic process can take;
2. An *initial state* probability distribution, i.e. $\pi = \{P(S_i \mid t = 0), S_i \in S\}$, where t is a discrete time index;
3. A probability distribution that characterizes the allowed transitions between states, that is $\mathbf{a} = \{P(S_j \text{ at time } t \mid S_i \text{ at time } t-1), S_i \in S, S_j \in S\}$ where the *transition probabilities* a_{ij} are assumed to be independent of time t;
4. An *observation* or *feature* space F, which is a discrete or continuous universe of all possible observable events (usually a subset of $\mathbb{R}^d$, where d is the dimensionality of the observations);
5. A set of probability distributions (referred to as *emission* or *output* probabilities) that describes the statistical properties of the observations for each state of the model: $\mathbf{b_x} = \{b_i(\mathbf{x}) = P(\mathbf{x} \mid S_i), S_i \in S, \mathbf{x} \in F\}$.

Figure 2.5 shows a simple, discrete HMM with three states. HMMs represent a learning paradigm, in the sense that examples of the event that is to be modeled can be collected and used in conjunction with a training algorithm in order to learn proper estimates of π, $\mathbf{a}$ and $\mathbf{b_x}$. The most popular of such algorithms are the forward-backward (or Baum-Welch) and the Viterbi algorithms [251]. Whenever continuous emission probabilities are considered, both of them are based on the general maximum-likelihood criterion, i.e. they aim at maximizing the probability of the samples given the model at hand. In particular, the Viterbi algorithm concentrates only on the most alike path throughout all the possible sequences of states in the model. These algorithms belong to the class of unsupervised learning techniques, since they perform unsupervised parameter estimation of the probability distributions without requiring any prior labeling of individual observations (within the sequences used for training) as belonging to specific states. Although HMMs fall in the category of *generative* models, once training has been accomplished the HMM can be used for decoding, or recognition, of sequences having arbitrary length.

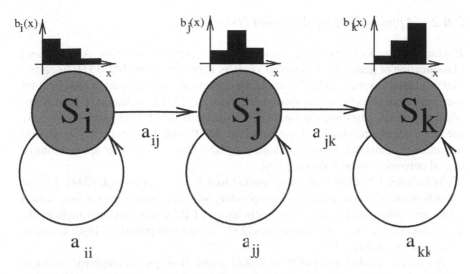

Fig. 2.5 A left-to-right, discrete hidden Markov model: can it be interpreted as a particular case of dynamic Bayesian network?

The interpretation of HMMs as dynamic Bayesian networks goes, roughly, as follows. Nodes are inserted in the graph for each hidden state q_t at time t, and for the emission probabilities associated with these states. Causal, directed edges are inserted between each pair of adjacent states (meaning that the state at time $t + 1$ is independent of all non-descendant states given the state at time t, due to the Markov assumption), and between each state and the corresponding emission (the latter being a function of the former, independently from the specific time t as well as from any other state in the model). The distribution of initial and transition probabilities, and the emission distributions as well, provide us with the quantities needed in order to fill the CPTs (depending only on the parents of a given node in the graph).

There is only one caveat, which can be qualitatively seen if we take a good look at Figure 2.5 (which, by the way, is representative of the way HMMs are thought of by practitioners in the real world). In general, the topology of the transitions renders the graph cyclic. In other words, straightforward interpretation of HMMs in terms of BNs seems to be somewhat fallacious. Actually, we can simply escape this pitfall by 'unfolding' in time the graphical structure, i.e., by replicating the vertex of each state variable for as many time steps as the hidden process is going to remain in that very state. Each pair of such duplicate nodes is then connected with a forward, directed edge which is a replica of the original, recursive self-transition of the state onto itself. This approach may remind us of the 'unfolding in time' algorithm used for training recurrent neural networks [127] via backpropagation through time over sequences [133]. The unfolding is also closely related to the 'trellis' structure used by the forward-backward and Viterbi algorithms for standard HMMs. Only problem is that, in so doing, sequences of different length will entail graphs of different depth, that is, different BNs. This renders inference and learning even harder.

2.6.2   Applications of Bayesian Networks

Bayesian networks found application in a wide range of scenarios. Analysis and classification of gene expression data via BNs is presented in [102, 131]. In particular, variations to the standard learning algorithms are investigated in [131] within the framework of genetic pattern classification. In [231] Bayesian inference is applied in order to carry out phylogeny reconstruction, developing previous work by the same authors relying on Bayesian networks. Probabilistic expert systems [53] are another significant area of application of BNs [190, 225]. Other applications of causal networks can be found in [286].

Aggregated BN classifiers were applied to a face detection task [244], i.e. the identification of visual patterns corresponding with the presence of a face within an image. Other image processing applications of Bayesian networks include object recognition in sequences of images [232], image interpretation [184], semantic image understanding [201].

The recent developments of Web-related applications put an emphasis on document analysis and automatic text categorization, as well as on information extraction. Hierarchical text classification via Bayesian networks is investigated, for instance, in [41]. Text classification via 'very large Bayesian multinets' is presented in [169]. A Bayesian network model for semi-structured document classification is outlined in [63].

Dynamic Bayesian networks, either in the form of Kalman filters or hidden Markov models, have been (implicitly or explicitly) applied to broad areas of signal processing [252], automatic speech recognition and speaker identification [58], handwritten text recognition [144], and a number of tasks in genomics and proteomics [74, 12].

2.6.3   Strengths and Weaknesses of Bayesian Networks

This variety of successful applications of Bayesian networks—along with the presentation we gave of the paradigm in the previous sections—points out that Bayesian networks are an interesting, powerful tool. BNs and their parameter/structure learning algorithms are sound, yet simple from the conceptual and the implementation viewpoints. Good results are obtained in even severe, real-world tasks. Most machine learning paradigms (such as neural networks [127] and support vector machines [310]) tend to behave like 'black boxes'. In fact, it is difficult to incorporate prior knowledge within them, and it is basically impossible to come out with real explanations/understanding of the possible meaning (if any) their parameters (as learned from the data) might have. On the contrary, Bayesian networks may be designed according to the knowledge of experts in a natural way (to this end, causal networks are the brightest instance). Correspondingly, after training they may be given an interpretation in terms of causal relations among the variables. Nonetheless, there are some limitations of BNs that have to be pointed out.

One major caveat concerns the graphical structure of a BN. If the independencies among the variables are fixed in advance (by intervention of a human expert) then the graph underlying the Bayesian network is sound, and no structure learning is needed. If, on the other end, structure learning is required, then the computational burden may become prohibitive on large datasets, approximate learning algorithms are unlike to find close-to-optimal solutions, and an interpretation of the corresponding probabilistic relations by experts may be difficult. Feasibility of the structure learning algorithm is one of the motivations that will guide us in the development of a novel, more effective graphical model representation in Chapter 4.

Another intrinsic constraint of BNs (and, broadly speaking, of all conventional graphical models) is that, in their classic formulation, they only deal with random vectors and not with *random graphs*, i.e. they cannot cope with structured data patterns which may be variable in size or in structure. For instance, in several bioinformatics applications the data are in the form of labeled graphs that represent molecules. Individual molecules have a different number of atoms (i.e., a variable vertex set), and a different chemical structure (i.e., edges in the graph are not fixed). Beginners' enthusiasm on plausible application of Bayesian networks to such scenarios might be smothered once they realize that the graphical nature of the machine does not entail its capability of dealing with data which are graphs themselves. Practitioners may well resort, to this end, to the many alternative machine learning paradigms which are suitable for learning over graphs [149, 285, 124, 298].

Chapter 3
Markov Random Fields

> *"[W]e avoid the gravest difficulties when, giving up the attempt to frame hypotheses concerning the constitution of material bodies, we pursue statistical inquiries as a branch of rational mechanics."*
>
> J. Willard Gibbs, 1902 [111]

3.1 Introduction

Let's give Bayesian networks a break, and let us go back to our favorite topic, namely soccer. Suppose you want to develop a probabilistic model of the ranking of your team in the domestic soccer league championship at any given time t throughout the current season. In this setup, it is reasonable to assume that t is a discrete time index, denoting t-th game in the season and ranging from $t = 1$ (first match of the tournament) to $t = T$ (season finale). Assuming the championship is organized as a round-robin tournament among N teams, then $T = 2(N - 1)$. The ranking of your team at time $t + 1$ is likely to change with a certain probability distribution which (i) accounts for the randomness of the results at the end of the corresponding matchday, and (ii) depends on the ranking at time t. For instance, it is definitely unlikely that the team moves from the bottom of the table to its top after, say, eighteenth matchday. We can describe the ranking at time t as the 'state' (i.e., outcome) of a random variable whose distribution changes over time, depending on previous states at time $t = 1, \ldots, t - 1$. We are familiar with the idea of a time-indexed variable X_t already, since we introduced it in Section 2.6.1 for presenting dynamic Bayesian networks. In statistics, this is an instance of what is known as a random (or stochastic) process [71]. Roughly speaking, we can state that a stochastic process is a set (thought of as a sequence, or time series) or time indexed random variables, i.e. $\{X_t \mid t \in \mathcal{T}\}$. The set $\mathcal{T}$ is the time domain, and may be discrete (e.g., $\mathcal{T} = \{1, \ldots, T\}$, as in the soccer example) or continuous-valued (e.g., $\mathcal{T} \subset \mathbb{R}$). The variable X_t may itself be discrete or continuous in nature, too. Let us focus on discrete-time processes. If X_{t+1} is independent of X_t for any $t \in \mathcal{T}$, then the process is said to be memoryless. In most, interesting cases there are complex statistical dependencies among the distributions of the variables at different times. If, for a certain integer $n > 0$, X_t depends on $X_{t-n}, \ldots, X_{t-1}$, and it is independent of $X_1, \ldots, X_{t-n-1}$ given $X_{t-n}, \ldots, X_{t-1}$, then the Markov property holds and the process takes the name of Markov chain. This clearly reminds us of Bayesian networks in general, and of hidden Markov models

A. Freno and E. Trentin: Hybrid Random Fields, ISRL 15, pp. 43–68.
springerlink.com　　　© Springer-Verlag Berlin Heidelberg 2011

in particular. Actually, as we saw, the hidden part of a hidden Markov model is a Markov chain. A Markov chain model could be reasonable to assume in the soccer championship ranking example, although it would be likely to lose some relevant statistical information on the long-term trend of the teams in the tournament (e.g., on the final matchday your team could still be on top, one point ahead of the runner-up, but knowing that it lost the last five games in a row whilst its rival won all of them would possibly change your mind on the likelihood of the final outcomes of the season).

We can generalize the notion of stochastic process as follows. Let us assume the set $\mathcal{T}$ of time indexes for the collection of random variables $\{X_t \mid t \in \mathcal{T}\}$ is now a set of d-dimensional integer or real-valued vectors. For instance, each $t \in \mathcal{T}$ might represent a specific location in a 2-dimensional space (like an entry on a grid, whose value is drawn from the distribution of the random variable $X_{(x_1,x_2)}$), or a point in 3-dim space at a given time. If $\mathcal{T}$ is a topological space [16] (or, a compact subset of this space), such as the Euclidean space, then $\{X_t \mid t \in \mathcal{T}\}$ is said to be a *random field* [2]. It ends up that stochastic processes are a particular, relevant instance of random fields.

The following example will be particularly fruitful. Let $X_{(\lambda,\phi)}$ represent the annual average temperature (measured throughout the last year using, say, the Celsius scale) at longitude λ and latitude ϕ on the surface of planet Earth. We can stick to the discrete setup by rounding the measurements to the closest integer Celsius degree, and by imagining that the coordinates (λ, ϕ) run over a countable set of pairs corresponding with the position of all the towns and villages over the world. Whenever we sample $X_{(\lambda,\phi)}$, the random value we obtain depends on several factors, such as: (*i*) the specific values of longitude and latitude (note that neighboring locations are likely to exhibit similar temperatures), e.g. the coordinates fall within a subtropical climate region; (*ii*) specific geological characteristics at (λ, ϕ) (which, in turn, are possibly shared by cities in the immediate surroundings); (*iii*) the passage of fronts of cold/warm air masses (e.g., according to fluid dynamics phenomena in the troposphere) which, once again, have just transited over one or more neighboring locations. In other words, $X_{(\lambda,\phi)}$ is highly correlated to the temperature of locations in the neighborhood of (λ, ϕ). On the other way around, as distance from (λ, ϕ) increases, this phenomenon of direct, mutual affection between the corresponding temperatures tends to disappear. In particular, the cold air masses passing over a city will affect the temperature at a distant location only by passing by (and, affecting the temperature of) intermediate towns along their journey through the atmosphere. As a consequence, it is 'reasonable' to assume that the probability distribution of $X_{(\lambda,\phi)}$ given the temperature of all other locations on the face of Earth reduces to the distribution of $X_{(\lambda,\phi)}$ given the temperature measured over the *neighbors* of (λ, ϕ). In summary, the Markov property holds, and the random field $\{X_{(\lambda,\phi)} \mid -180° \le \lambda \le 180°, -90° \le \phi \le 90°\}$ is then said to be a *Markov random field* (MRF, also referred to as *Markov network*). In the example given, if we represent the temperature of each town with a vertex $X_{(\lambda,\phi)}$ in a graph, and we insert edges between nodes corresponding to 'neighboring' locations (that is, towns whose climates exercise a direct, mutual influence on each other), we can look at the

MRF as a graphical model. At this qualitative level of description, it should be easy to intuit that the statistical relationship among neighboring variables are symmetric, that is, an *undirected* graph is now opportune, in contrast with the directed graphical structure occurring in the definition of Bayesian networks. Also, the graph may result to be cyclic (weather conditions influence neighboring locations which, in turn, affect intermediate locations etc., eventually leading to a complete trip around the world and back to the original place). In this chapter, it will be seen that the abstract concept of 'neighborhood' is formalized by means of the (undirected) notion of *adjacency* (please refer to Appendix B.1 for a review of these definitions in graph theory).

The example of toy meteorology is also of interest from the topological point of view. In fact, it points out that the index universe $\mathcal{J}$ occurring in the definition of a random field may as well be a *manifold* [135]. The surface of Earth (our planet being 3-dimensional) can indeed be roughly mapped onto a collection (mathematically speaking, an *atlas*) of local, 2-dim *charts*, preserving the topology (including the notion of neighborhood underlying the definition of Markov networks).

A different way to approach the qualitative idea of Markov random field (and to introduce other notions, required to arrive at a formalization of MRFs) is rooted in physics, specifically in statistical mechanics [145, 45]. Actually, most authors trace the history of MRFs back to the so-called Ising model [194]. This model describes a substance in terms of a lattice (i.e., a graph) of discrete, interacting units called spins. Each spin may assume as value either 1 or -1,[1] according to an *energy function* which accounts for local interactions between a spin and its neighbors in the graph. The model aims at explaining phase transitions in the corresponding substance. From a statistical viewpoint, the probability of the value the spins may assume at a given time is interpreted as a function of the energy. More precisely, a Boltzmann-Gibbs distribution proportional to $\exp(-\beta E)$ is assumed, where E is the energy of the system, and β is (proportional to) its inverse temperature.

Each spin in the system affects its neighbors according to the mutual interaction between the corresponding magnetic fields. Roughly speaking, a *field* is a function (e.g., a scalar or vector function) defined over a domain which contains the spatial coordinates of any locations in the system. This mathematical notion is of the utmost relevance to physics, where fields are used to describe physical quantities as functions of space and time. Examples are the magnetic field and Newton gravitational field, the strength of which is known to vanish with the square of the distance from the source. This is a general tendency of several fields that have been observed in physics, and it is representative of the idea of focusing on local interactions in a neighborhood of the source, in spite of the fact that, from a mathematical point of view, the field extends all over its universe of definition. We will see that Markov random fields exploit these ideas. They assume a total, field-like energy (that theoretically extends over the system at large, but that can be suitably modeled in terms of symmetric interactions within local neighborhoods) according to an exponential probability distribution (being, in turn, a function of local energies).

---

[1] In physics, the alternative values $-\frac{1}{2}$ and $\frac{1}{2}$ are commonly used.

Vector fields, such as the gravitational or magnetic fields, are indeed defined for a specific *energy function*, known as the *Hamiltonian*. The overall energy of the system as a whole is expected to be the value of the Hamiltonian. In mechanics, at any given location within the system, a certain potential energy is defined. It is basically the amount of energy needed for moving a point in the field from a given location to another. For instance, this book you are reading right now has a certain potential energy under the influence of the gravitational field (just let the book fall to the ground, if you feel skeptical). The Hamiltonian (total energy) is the integral over the system of kinetic and potential energies. If a field is defined over a discrete set of steady locations, or units (i.e., having null kinetics), then the Hamiltonian reduces to the sum, extended to all locations, of unit-defined *potential functions*. These are scalar functions which express the potential energy at a local level. No surprise Markov random fields do the same, defining local neighborhood-specific potential functions. In physics a force field is, in fact, the negative gradient of its potential function[2]. The reader will be delighted to find out that, in a research conducted at the University of California at Berkeley, potential functions were actually used for modeling the flow of play in soccer games [35].

Now, let us assume we wish to calculate the probability of a random field being in a specific state. For instance, the state of the system could be the set of $+1/-1$ values associated with the spins in the Ising model (in abstract, the state could be any assignment of values to the variables in the random field). Some important results from physics allow us to express this probability in a compact form, relying on the quantities introduced so far (potential, temperature, energy). In particular, a *partition function* $Z(\beta)$ (β being proportional to the inverse temperature of the system) may be used to this end, whenever the system is in thermodynamic equilibrium [145, 45]. It is defined as $Z(\beta) = \sum_\sigma \exp\{-\beta H(\sigma)\}$, where the sum is extended to all possible states of the system and $H(\sigma)$ is the Hamiltonian, i.e. the overall energy of the system in state σ. The Hamiltonian can be expressed in terms of potential functions $\varphi(\xi)$, evaluated (in state σ) over all symmetric 'neighborhoods' ξ in the system, as $H(\sigma) = \sum_\xi \varphi(\xi)$. In so doing, the partition function may be rewritten in the following form [45]:

$$Z(\beta) = \sum_\sigma \exp\{-\beta \sum_\xi \varphi(\xi)\} \tag{3.1}$$

The probability of a specific state σ^* is now obtained as [145, 45]:

$$P(\sigma^*) = \frac{1}{Z(\beta)} \exp\{-\beta \sum_\xi \varphi(\xi)\} \tag{3.2}$$

This is a significant result in statistical mechanics. Moreover, it turns out that MRFs express the joint probability of a set of random variables on a lattice according to an equation having similar form.

---

[2] Note that in mathematics a conservative vector field is often defined as the *positive* gradient of its potential function.

The formal definition of Markov random field as a probabilistic graphical model is given in Section 3.2, which explains how to derive a joint probability distribution from a MRF. Section 3.2.1 illustrates the statistical assumptions underlying the use of Markov networks. In particular, the Hammersley-Clifford theorem is stated, which constitutes the theoretical basis for a proper probabilistic interpretation of the model and of the joint density it encapsulates. Since the computation of the standard representation of joint probabilities in MRFs results to be intractable, Section 3.2.2 introduces a more efficient way of computing joint distributions, in the form of a pseudo-likelihood function. In order to compute the overall pseudo-likelihood, we need to rely on conditional probabilities of individual nodes given their Markov blankets. This topic is covered in Section 3.2.3.

Once the definitions underlying the very nature of MRFs are given, and a specific graphical structure is known, we need techniques for learning the parameters which characterize the probability distributions. In Section 3.3 we show how to learn the parameters of a Markov random field from a given dataset. Section 3.3.1 describes a gradient-based approach for the optimization of the *weights* that are involved in the formal definition of the potential functions associated with a MRF. Section 3.3.2 provides an algorithm for finding all maximal cliques in an undirected graph: performing this task is in fact necessary in order to initialize the structure of MRFs (i.e., to the end of defining the model parameters).

As we saw in the previous chapter, the graphical structure of Bayesian networks is often fixed by human experts according to a certain prior knowledge of the application domain, expressing a causal relationship between pairs of variables. Nonetheless, a structure learning strategy ends up to be necessary in the general case. Although in several cases the graphical structure of MRFs is just a more or less natural consequence of the specific application domain (the literature on MRFs mostly assumes the graph is known, as in the toy examples above), we cannot underestimate the necessity for suitable techniques that can infer the topology of the graph at large. The issue is investigated in Section 3.4, where an algorithm for structure learning from data is discussed. Both parameter learning and structure learning prove to be much more complex in MRFs then they used to be in BNs.

Finally, Section 3.5 offers concluding remarks, including a concise survey of generalizations of the basic notion of Markov network, a list of their points of strength and drawbacks, and a selection of representative applications over a wide range of scenarios.

Although the history of Markov networks can be traced back to 1925, with the 1-dimensional Ising model, it is likely that the first systematic treatment of what is known, to date, as a MRF was accomplished in 1980 by Ross Kindermann and James Laurie Snell [167]. Statistical mechanics approaches (namely, simulated annealing) properly related to MRFs were first proposed in the mid Eighties by Stuart Geman and Donald Geman [106, 281], who applied the model to the restoration of images. Application of Gibbs sampling and Markov chain Monte Carlo techniques were first presented in 1974, thanks to the seminal work by John Moussouris [221], and systematically discussed in the Nineties by Julian Besag and other authors (see, for instance, [26]). More or less recent, comprehensive reviews of MRFs and their applications can be found, just to mention a few, in [265], [47], and [195].

3.2 Representation of Probabilities

Markov random fields [121, 167, 282] are used to represent joint probability distributions underlying sets of random variables. A Markov random field is composed of an *undirected graph* and a set of *potential functions*. Each node in the graph represents a random variable, and for each maximal clique[3] $\mathcal{C}$ in the graph we have a corresponding potential function $\varphi_\mathcal{C}$. Each potential function $\varphi_\mathcal{C}$ takes as argument the *state* of clique $\mathcal{C}$ in the graph and returns as value a non-negative real number. The state of a clique $\mathcal{C}$ in a Markov random field is nothing but a specific realization of the variables in $\mathcal{C}$, i.e. an event described by a specific configuration of the values of those variables. Figure 3.1 exemplifies the graphical component of a Markov random field.

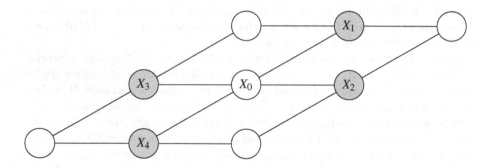

Fig. 3.1 The graphical component of a Markov random field. The node X_0 is contained in four different maximal cliques, i.e. $\mathcal{C}_1 = \{X_0, X_1\}$, $\mathcal{C}_2 = \{X_0, X_2\}$, $\mathcal{C}_3 = \{X_0, X_3\}$, and $\mathcal{C}_4 = \{X_0, X_4\}$. Shaded nodes denote the *neighborhood* of X_0 (i.e. the set containing all the nodes that are directly linked to X_0), given by $\mathcal{N}(X_0) = \bigcup_{i=1}^{4} \mathcal{C}_i \setminus \{X_0\} = \{X_1, X_2, X_3, X_4\}$.

Consider a set $\mathbf{X}$ of random variables $X_1, \ldots, X_n$, a graph $\mathcal{G}$ whose nodes are the variables in $\mathbf{X}$, and the set γ of all maximal cliques in $\mathcal{G}$. Given the potential functions for the cliques in γ, the joint probability distribution of $\mathbf{X}$ is computed as follows:

$$P(\mathbf{x}) = \frac{1}{Z} \prod_{\mathcal{C} \in \gamma} \varphi_\mathcal{C}(\mathbf{x}_\mathcal{C}) \tag{3.3}$$

where Z is a normalization factor called 'partition function' and $\mathbf{x}_\mathcal{C}$ is the state of clique $\mathcal{C}$ as this state is determined by $\mathbf{x}$. If $\mathcal{X}$ is the set of all possible configurations of the values of the variables in $\mathbf{X}$, the partition function is given by:

---

[3] See Appendix B.1 for the definitions of clique and maximal clique in undirected graphs.

$$Z = \sum_{\mathbf{x} \in \mathcal{X}} \prod_{\mathcal{C} \in \gamma} \varphi_{\mathcal{C}}(\mathbf{x}_{\mathcal{C}}) \tag{3.4}$$

Since the cardinality of $\mathcal{X}$ is given by $|\mathcal{X}| = \prod_{i=1}^{n} |\mathcal{D}_i|$, where $\mathcal{D}_i$ denotes the domain of variable X_i, it is clear that computing Z requires to sum over a number of states that grows exponentially with the number of variables in $\mathbf{X}$. Therefore, if we stick to the probability function specified in Equation 3.3, estimating the likelihood of a state $\mathbf{x}$ for a given Markov random field remains intractable. A solution to this problem, first suggested in [25], will be detailed in Section 3.2.2.

The potential functions are represented as exponential functions. This choice involves exploiting the notion of *feature functions*. For each state $\mathbf{x}_{\mathcal{C}}$ of each maximal clique $\mathcal{C}$, we introduce a feature function $f_{\mathbf{x}_{\mathcal{C}}}$ such that:

$$f_{\mathbf{x}_{\mathcal{C}}}(\mathbf{X}_{\mathcal{C}}) = \begin{cases} 1 \text{ if } \mathbf{X}_{\mathcal{C}} = \mathbf{x}_{\mathcal{C}} \\ 0 \text{ otherwise} \end{cases} \tag{3.5}$$

In other words, a feature function returns 1 if the respective clique is in the state corresponding to that feature function, while it returns 0 if the clique is in any other state. To each feature function f_j we attach a real-valued weight w_j. Given the feature functions and their weights, each potential function $\varphi_{\mathcal{C}}$ takes the following form:

$$\varphi_{\mathcal{C}}(\mathbf{x}_{\mathcal{C}}) = \exp\left(\sum_{j=1}^{m_{\mathcal{C}}} w_j f_j(\mathbf{x}_{\mathcal{C}})\right) \tag{3.6}$$

where $m_{\mathcal{C}}$ is the number of features defined for clique $\mathcal{C}$. This leads to the following representation of the joint probability distribution:

$$\begin{aligned} P(\mathbf{x}) &= \frac{1}{Z} \prod_{\mathcal{C} \in \gamma} \exp\left(\sum_{j=1}^{m_{\mathcal{C}}} w_j f_j(\mathbf{x}_{\mathcal{C}})\right) \\ &= \frac{1}{Z} \exp\left(\sum_{\mathcal{C} \in \gamma} \sum_{j=1}^{m_{\mathcal{C}}} w_j f_j(\mathbf{x}_{\mathcal{C}})\right) \end{aligned} \tag{3.7}$$

Concerning the weights, they can be learned from data so as to maximize the probability of the data given the Markov random field. This problem will be addressed in section 3.3.

3.2.1 The Hammersley-Clifford Theorem

The kind of probability distribution specified in Equation 3.7 is usually referred to as a *Gibbs* (or *Boltzmann*) *distribution*. At this point, the reader may well wonder why we should choose to model joint probabilities by means of Gibbs distributions, and in particular why we should choose to factorize our Gibbs distribution based on the maximal cliques of a certain (undirected) graph. The key to understanding the connection between the Gibbs distribution and the graphical structure of Markov random fields is provided by the Hammersley-Clifford theorem [24].

Before stating the theorem, let us first define the notion of *(undirected) separation*:

Definition 3.1. Consider an undirected graph $\mathcal{G} = (\mathcal{V}, \mathcal{E})$. If $\mathcal{A}$, $\mathcal{B}$, and $\mathcal{S}$ are three disjoint subsets of $\mathcal{V}$, then $\mathcal{S}$ *separates* $\mathcal{A}$ from $\mathcal{B}$ if, for any node X_i in $\mathcal{A}$ and any node X_j in $\mathcal{B}$, every possible path connecting X_i to X_j contains at least a node X_k such that $X_k \in \mathcal{S}$.

Given the concept of separation, the Hammersley-Clifford theorem can be stated as follows:

Theorem 3.1. *Given a random vector* $\mathbf{X} = (X_1, \ldots, X_n)$, *if* $\mathcal{G} = (\mathcal{V}, \mathcal{E})$ *is an undirected graph such that* $\mathcal{V} = \{X_1, \ldots, X_n\}$ *and P is a probability distribution over* $\mathbf{X}$ *such that* $P(\mathbf{x}) > 0$ *for any realization* $\mathbf{x}$ *of* $\mathbf{X}$, *then the following conditions are equivalent:*

1. *$P(\mathbf{X})$ is a Gibbs distribution which factorizes according to the maximal cliques in $\mathcal{G}$;*
2. *If $\mathcal{N}(X_i)$ is the set of neighbors of X_i in $\mathcal{G}$, then $P(X_i|\mathbf{X} \setminus \{X_i\}) = P(X_i \mid \mathcal{N}(X_i))$ (local Markov property);*
3. *If $\mathcal{A}$, $\mathcal{B}$, and $\mathcal{S}$ are three disjoint subsets of $\mathbf{X}$, and $\mathcal{S}$ separates $\mathcal{A}$ from $\mathcal{B}$, then $P(\mathcal{A} \mid \mathcal{B} \cup \mathcal{S}) = P(\mathcal{A} \mid \mathcal{S})$ (global Markov property).*

Proof. See e.g. [24].   □

In other words, given a certain domain of random variables, the Hammersley-Clifford theorem entails that assuming either that the domain can be modeled by a Gibbs distribution or that the domain satisfies the local or the global Markov property means assuming one and the same thing. Therefore, using a Markov random field to model a certain probability distribution is justified exactly when any one of the three conditions specified in the theorem is satisified by the distribution at hand. That is to say, when we use Markov random fields we are in fact making any one of the three assumptions proved to be equivalent by Hammersley and Clifford.

The *positivity condition* on the probability distribution, i.e. the requirement that $P(\mathbf{x}) > 0$, is strictly necessary in order for the theorem to hold. This condition played a singular role in the history of the Hammersley-Clifford theorem.[4] The theorem was first proved by John Hammersley and Peter Clifford in 1971, assuming the positivity condition. However, the authors of the proof felt that this assumption was an undesired feature of their result. The drawback of the positivity condition is that it does not allow to apply Markov random fields to any domain featuring *forbidden states*,

---

[4] The historical information contained in this paragraph are drawn from Hammersley's remarks on [24], contained in an additional section of Besag's paper featuring a critical discussion by different authors.

i.e. to any system such that some constraints prevent it from assuming certain states. This limitation ruled out the possibility of modeling some problems in statistical mechanics through Markov random fields. For this reason, Hammersley and Clifford postponed the publication of the result, in the hope of proving the theorem without assuming the positivity condition. In the following years, Julian Besag worked out a more elegant proof of the theorem than Hammersley and Clifford's original proof, which was revealed to be unduly complex, and Besag's result was published [24]. Of course, the positivity condition was assumed by Besag too in his proof of the theorem. Finally, the last chapter in the history of the Hammersley-Clifford theorem was written by one of Hammersley's graduate students at Oxford, namely John Moussouris, who showed (by means of a counter-example) that the positivity of the probability distribution was a strictly necessary condition of the theorem [221].

3.2.2 A Pseudo-Likelihood Measure

Now, we have to face an important problem. Unfortunately, computing the partition function specified in Equation 3.4 requires summing over a number of states which grows exponentially with the number of variables. This is clearly unfeasible in most real-world settings. A workaround for this difficulty was proposed by Julian Besag [25]. Besag's idea is to replace the likelihood function in Equation 3.3 with the following *pseudo-likelihood* measure:

$$P^*(\mathbf{X} = \mathbf{x}) = \prod_{i=1}^{n} P(X_i = x_i \mid \mathbf{X} \setminus \{X_i\} = \mathbf{x} \setminus \{x_i\}) \tag{3.8}$$

In other words, the idea underlying the pseudo-likelihood approach is to factorize the joint distribution of the random vector $\mathbf{X}$ as the product of the distributions of each variable X_i given the values of all the variables X_j such that $j \neq i$. Although the pseudo-likelihood function does not satisfy the axioms of probability theory, both theoretical and experimental evidence has been put forward in the relevant literature for the plausibility of employing pseudo-likelihood as an alternative to a proper joint probability measure [25, 256, 228, 95]. One convenient property of the pseudo-likelihood function with respect to graphical models is that, as defined by Equation 3.8, it reduces to the following function:

$$P^*(\mathbf{X} = \mathbf{x}) = \prod_{i=1}^{n} P(X_i = x_i \mid mb(X_i)) \tag{3.9}$$

where $mb(X_i)$ refers to the state of the Markov blanket of X_i. The general notion of Markov blanket was defined in Chapter 2 (see Definition 2.4). Now, as a result of Theorem 3.1, a Markov blanket for X_i is provided by $\mathcal{N}(X_i)$, i.e. by the neighborhood of X_i. Since the set $\mathcal{N}(X_i)$ is unique, when we refer to the Markov blanket

$\mathcal{MB}(X_i)$ of variable X_i we actually mean $\mathcal{N}(X_i)$. The clear advantage of the pseudo-likelihood function over the standard likelihood measure is that the former dispenses with the partition function embedded in the latter, and thereby it drops an exponentially growing amount of calculations.

3.2.3 Markov Blankets

In order to compute the pseudo-likelihood of a Markov random field, we must be able to compute the conditional probability of each node given its Markov blanket, i.e. given its neighborhood. Before explaining how we compute that probability, we need to introduce the following notation. If $\mathbf{x}$ is a vector of values for the variables in a given set $\mathbf{X}$, and x_i is a specific value of a variable X_i contained in $\mathbf{X}$, then $f(\mathbf{x}, x_i)$ denotes the value of the feature function $f(\mathbf{x}^*)$, where the vector $\mathbf{x}^*$ is the same as $\mathbf{x}$ except that we force X_i to assume value x_i. Given this notation, the conditional probability distribution of a variable X given that its Markov blanket $\mathcal{MB}(X)$ is in state $mb(X)$ is computed as follows, for any value x of X:

$$P(x \mid mb(X)) = \frac{\exp\left(\sum_{\mathcal{C} \in \gamma_X} \sum_{j=1}^{m_\mathcal{C}} w_j f_j(\mathbf{x}_\mathcal{C})\right)}{\sum_{i=1}^{|\mathcal{D}_X|} \exp\left(\sum_{\mathcal{C} \in \gamma_X} \sum_{j=1}^{m_\mathcal{C}} w_j f_j(\mathbf{x}_\mathcal{C}, x_i)\right)} \tag{3.10}$$

where $\gamma_X = \{\mathcal{C} : \mathcal{C} \in \gamma \wedge X \in \mathcal{C}\}$ and $\mathcal{D}_X$ is the domain of X. The computation described in Equation 3.10 is justified as follows. First, we have that

$$P(x \mid mb(X)) = \frac{P(x, mb(X))}{P(mb(X))} \tag{3.11}$$

Since $P(mb(X))$ remains constant for any value of X, we can ignore that term, obtaining the following equation:

$$P(x \mid mb(X)) \propto P(x, mb(X)) \tag{3.12}$$

Now, if $\mathcal{X}$ is the set of all possible joint states of the variables in $\mathcal{G}$, consider the set $\mathcal{X}^* \subset \mathcal{X}$ of those joint states $\mathbf{x}$ such that $X = x$ and $\mathcal{MB}(X) = mb(X)$. Clearly,

$$
\begin{aligned}
P(x, mb(X)) &= \sum_{\mathbf{x} \in \mathcal{X}^*} P(\mathbf{x}) \\
&= \sum_{\mathbf{x} \in \mathcal{X}^*} \frac{1}{Z} \prod_{\mathcal{C} \in \gamma} \varphi_\mathcal{C}(\mathbf{x}_\mathcal{C}) \\
&= \frac{1}{Z} \sum_{\mathbf{x} \in \mathcal{X}^*} \prod_{\mathcal{C} \in \gamma} \varphi_\mathcal{C}(\mathbf{x}_\mathcal{C}) \\
&\propto \sum_{\mathbf{x} \in \mathcal{X}^*} \prod_{\mathcal{C} \in \gamma} \varphi_\mathcal{C}(\mathbf{x}_\mathcal{C})
\end{aligned}
\tag{3.13}
$$

where the last step is justified by the fact that Z is constant for a given Markov random field. Since $\prod_{c \in \gamma_X} \varphi_c(\mathbf{x}_c)$ is constant for any $\mathbf{x}$ in $\mathcal{X}^*$, the last quantity specified in derivation 3.13 can be rewritten as follows:

$$\sum_{\mathbf{x} \in \mathcal{X}^*} \prod_{c \in \gamma} \varphi_c(\mathbf{x}_c) = \prod_{c \in \gamma_X} \varphi_c(\mathbf{x}_c) \sum_{\mathbf{x} \in \mathcal{X}^*} \prod_{c^* \notin \gamma_X} \varphi_{c^*}(\mathbf{x}_{c^*}) \tag{3.14}$$

In other words, we have that

$$P(x \mid mb(X)) \propto \prod_{c \in \gamma_X} \varphi_c(\mathbf{x}_c) \sum_{\mathbf{x} \in \mathcal{X}^*} \prod_{c^* \notin \gamma_X} \varphi_{c^*}(\mathbf{x}_{c^*}) \tag{3.15}$$

Given that the cliques that are not in γ_X do not contain node X, it follows that $\sum_{\mathbf{x} \in \mathcal{X}^*} \prod_{c^* \notin \gamma_X} \varphi_{c^*}(\mathbf{x}_{c^*})$ remains constant for any value x of X. Therefore, proportion 3.15 can be rewritten as

$$P(x \mid mb(X)) \propto \prod_{c \in \gamma_X} \varphi_c(\mathbf{x}_c) \tag{3.16}$$

Now, the value of $\prod_{c \in \gamma_X} \varphi_c(\mathbf{x}_c)$ is made explicit by the following equations:

$$\prod_{c \in \gamma_X} \varphi_c(\mathbf{x}_c) = \prod_{c \in \gamma_X} \exp\left(\sum_{j=1}^{m_c} w_j f_j(\mathbf{x}_c)\right)$$
$$= \exp\left(\sum_{c \in \gamma_X} \sum_{j=1}^{m_c} w_j f_j(\mathbf{x}_c)\right) \tag{3.17}$$

Therefore, the result of Equations 3.11–3.17 is the following:

$$P(x \mid mb(X)) \propto \exp\left(\sum_{c \in \gamma_X} \sum_{j=1}^{m_c} w_j f_j(\mathbf{x}_c)\right) \tag{3.18}$$

which completes the proof of Equation 3.10.

3.3 Parameter Learning

Parameter learning in Markov random fields is the problem of learning the weights of the feature functions from data. Since the pseudo-likelihood of a Markov random field given some specified data is a function of those weights, a straightforward way of dealing with this task is a maximum likelihood strategy aimed at finding the values of the weights that maximize the pseudo-likelihood function given the data.

3.3.1 Optimizing the Weights

Suppose we are given a dataset $\mathbf{D}$ containing m data points, where each data point $\mathbf{d}_j$ is a vector $(x_{1_j}, \ldots, x_{n_j})$, and a Markov random field h with weights $w_1, \ldots, w_l$. Pursuing a maximum likelihood strategy, we can then try to minimize the following objective function:

$$
\begin{aligned}
-\log P^*(\mathbf{D} \mid h) &= -\sum_{j=1}^{m} \log P^*(\mathbf{d}_j \mid h) \\
&= -\sum_{j=1}^{m} \sum_{i=1}^{n} \log P\left(X_i = x_{i_j} \mid mb_j(X_i)\right)
\end{aligned}
\tag{3.19}
$$

In other words, we want to minimize the negative logarithm of the pseudo-likelihood.

If we are able to compute the gradient of the objective function, which is indeed the case, we can then minimize that function by applying standard gradient-based optimization techniques [283]. Although several techniques may be used to this end (such as gradient descent), a convenient choice for the latter task is the limited-memory BFGS algorithm (L-BFGS), which is a quasi-Newton optimization method, first proposed in [196]. As compared to standard quasi-Newton methods, L-BFGS reduces the memory requirements of the optimization process. The algorithm does not compute explicitly the Hessian matrix, which would be quite expensive for problems involving a large number of variables. Therefore, the L-BFGS method is particularly well suited for large-scale optimization tasks.

In the following, we derive the gradient of the objective function through Equations 3.20–3.32, that show how to compute the partial derivatives of the objective function with respect to each weight w_k, whereas the results of the derivation will be given by Equations 3.33–3.35.

$$
\begin{aligned}
\frac{\partial}{\partial w_k} \left\{ -\sum_{j=1}^{m} \sum_{i=1}^{n} \log P\left(x_{i_j} \mid mb_j(X_i)\right) \right\} &= \\
= -\sum_{j=1}^{m} \sum_{i=1}^{n} \frac{\partial}{\partial w_k} \log P\left(x_{i_j} \mid mb_j(X_i)\right) & \\
= -\sum_{j=1}^{m} \sum_{X_i \in \mathcal{C}_{w_k}} \frac{\partial}{\partial w_k} \log P\left(x_{i_j} \mid mb_j(X_i)\right) & \\
= -\sum_{j=1}^{m} \sum_{X_i \in \mathcal{C}_{w_k}} \frac{1}{P\left(x_{i_j} \mid mb_j(X_i)\right)} \cdot \frac{\partial}{\partial w_k} P\left(x_{i_j} \mid mb_j(X_i)\right) &
\end{aligned}
\tag{3.20}
$$

where $\mathcal{C}_{w_k}$ denotes the clique corresponding to weight w_k. Based on Equation 3.10, we have:

$$\frac{\partial}{\partial w_k} P\left(x_{i_j} \mid mb_j(X_i)\right) = \frac{\partial}{\partial w_k} \left\{ \frac{\exp\left(\sum_{w_l \in \mathcal{W}_{X_i}} w_l f_l(\mathbf{d}_{j\mathcal{C}_{w_l}})\right)}{\sum_{i^*=1}^{|\mathcal{D}_{X_i}|} \exp\left(\sum_{w_l \in \mathcal{W}_{X_i}} w_l f_l(\mathbf{d}_{j\mathcal{C}_{w_l}}, x_{i^*})\right)} \right\} \quad (3.21)$$

where $\mathcal{W}_{X_i} = \{w : X_i \in \mathcal{C}_w\}$ and $\mathbf{d}_{j\mathcal{C}_{w_l}}$ is the state assumed by clique $\mathcal{C}_{w_l}$ in data point $\mathbf{d}_j$. We now use the following notation:

$$\alpha(w_k) = \exp\left(\sum_{w_l \in \mathcal{W}_{X_i}} w_l f_l(\mathbf{d}_{j\mathcal{C}_{w_l}})\right) \quad (3.22)$$

$$\beta(w_k) = \sum_{i^*=1}^{|\mathcal{D}_{X_i}|} \exp\left(\sum_{w_l \in \mathcal{W}_{X_i}} w_l f_l(\mathbf{d}_{j\mathcal{C}_{w_l}}, x_{i^*})\right) \quad (3.23)$$

Equation 3.21 can then be rewritten as follows:

$$\frac{\partial}{\partial w_k} P\left(x_{i_j} \mid mb_j(X_i)\right) = \frac{\partial}{\partial w_k} \left\{ \frac{\alpha(w_k)}{\beta(w_k)} \right\}$$

$$= \frac{1}{\beta(w_k)^2} \cdot \left(\frac{\partial \alpha(w_k)}{\partial w_k} \cdot \beta(w_k) - \alpha(w_k) \cdot \frac{\partial \beta(w_k)}{\partial w_k} \right) \quad (3.24)$$

Going back to Equation 3.20, we write:

$$-\sum_{j=1}^{m} \sum_{X_i \in \mathcal{C}_{w_k}} \frac{1}{P\left(x_{i_j} \mid mb_j(X_i)\right)} \cdot \frac{\partial}{\partial w_k} P\left(x_{i_j} \mid mb_j(X_i)\right) =$$

$$= -\sum_{j=1}^{m} \sum_{X_i \in \mathcal{C}_{w_k}} \frac{\beta(w_k)}{\alpha(w_k)} \cdot \frac{\partial}{\partial w_k} P\left(x_{i_j} \mid mb_j(X_i)\right) \quad (3.25)$$

$$= -\sum_{j=1}^{m} \sum_{X_i \in \mathcal{C}_{w_k}} \frac{1}{\alpha(w_k) \cdot \beta(w_k)} \cdot \left(\frac{\partial \alpha(w_k)}{\partial w_k} \cdot \beta(w_k) - \alpha(w_k) \cdot \frac{\partial \beta(w_k)}{\partial w_k} \right)$$

We now simplify $\frac{\partial}{\partial w_k} \alpha(w_k)$. First,

$$\frac{\partial}{\partial w_k} \alpha(w_k) = \frac{\partial}{\partial w_k} \exp\left(\sum_{w_l \in \mathcal{W}_{X_i}} w_l f_l(\mathbf{d}_{j\mathcal{C}_{w_l}})\right)$$

$$= \exp\left(\sum_{w_l \in \mathcal{W}_{X_i}} w_l f_l(\mathbf{d}_{j\mathcal{C}_{w_l}})\right) \cdot \sum_{w_l \in \mathcal{W}_{X_i}} \frac{\partial w_l f_l(\mathbf{d}_{j\mathcal{C}_{w_l}})}{\partial w_k} \quad (3.26)$$

Second,

$$\sum_{w_l \in \mathcal{W}_{X_i}} \frac{\partial w_l f_l(\mathbf{d}_{j_{\mathcal{C}_{w_l}}})}{\partial w_k} = \sum_{w_l \in \mathcal{W}_{X_i}} \left\{ \frac{\partial w_l}{\partial w_k} \cdot f_l(\mathbf{d}_{j_{\mathcal{C}_{w_l}}}) + w_l \cdot \frac{\partial f_l(\mathbf{d}_{j_{\mathcal{C}_{w_l}}})}{\partial w_k} \right\}$$

$$= \sum_{w_l \in \mathcal{W}_{X_i}} \frac{\partial w_l}{\partial w_k} \cdot f_l(\mathbf{d}_{j_{\mathcal{C}_{w_l}}}) \qquad (3.27)$$

$$= f_k(\mathbf{d}_{j_{\mathcal{C}_{w_l}}})$$

Therefore,

$$\frac{\partial}{\partial w_k} \alpha(w_k) = \exp\left(\sum_{w_l \in \mathcal{W}_{X_i}} w_l f_l(\mathbf{d}_{j_{\mathcal{C}_{w_l}}}) \right) \cdot f_k(\mathbf{d}_{j_{\mathcal{C}_{w_l}}}) \qquad (3.28)$$

$$= \alpha(w_k) \cdot f_k(\mathbf{d}_{j_{\mathcal{C}_{w_l}}})$$

Similarly, we simplify $\frac{\partial}{\partial w_k} \beta(w_k)$. First,

$$\frac{\partial}{\partial w_k} \beta(w_k) = \frac{\partial}{\partial w_k} \sum_{i^*=1}^{|\mathcal{D}_{X_i}|} \exp\left(\sum_{w_l \in \mathcal{W}_{X_i}} w_l f_l(\mathbf{d}_{j_{\mathcal{C}_{w_l}}}, x_{i^*}) \right)$$

$$= \sum_{i^*=1}^{|\mathcal{D}_{X_i}|} \frac{\partial \exp\left(\sum_{w_l \in \mathcal{W}_{X_i}} w_l f_l(\mathbf{d}_{j_{\mathcal{C}_{w_l}}}, x_{i^*}) \right)}{\partial w_k}$$

$$= \sum_{i^*=1}^{|\mathcal{D}_{X_i}|} \exp\left(\sum_{w_l \in \mathcal{W}_{X_i}} w_l f_l(\mathbf{d}_{j_{\mathcal{C}_{w_l}}}, x_{i^*}) \right) \cdot \sum_{w_l \in \mathcal{W}_{X_i}} \frac{\partial w_l f_l(\mathbf{d}_{j_{\mathcal{C}_{w_l}}}, x_{i^*})}{\partial w_k}$$

$$(3.29)$$

Now,

$$\sum_{w_l \in \mathcal{W}_{X_i}} \frac{\partial w_l f_l(\mathbf{d}_{j_{\mathcal{C}_{w_l}}}, x_{i^*})}{\partial w_k} = \sum_{w_l \in \mathcal{W}_{X_i}} \left\{ \frac{\partial w_l}{\partial w_k} \cdot f_l(\mathbf{d}_{j_{\mathcal{C}_{w_l}}}, x_{i^*}) + w_l \cdot \frac{\partial f_l(\mathbf{d}_{j_{\mathcal{C}_{w_l}}}, x_{i^*})}{\partial w_k} \right\}$$

$$= \sum_{w_l \in \mathcal{W}_{X_i}} \frac{\partial w_l}{\partial w_k} \cdot f_l(\mathbf{d}_{j_{\mathcal{C}_{w_l}}}, x_{i^*})$$

$$= f_k(\mathbf{d}_{j_{\mathcal{C}_{w_l}}}, x_{i^*})$$

$$(3.30)$$

Therefore,

$$\frac{\partial \beta(w_k)}{\partial w_k} = \sum_{i^*=1}^{|\mathcal{D}_{X_i}|} \exp\left(\sum_{w_l \in \mathcal{W}_{X_i}} w_l f_l(\mathbf{d}_{j_{\mathcal{C}_{w_l}}}, x_{i^*}) \right) \cdot f_k(\mathbf{d}_{j_{\mathcal{C}_{w_l}}}, x_{i^*}) \qquad (3.31)$$

Equation 3.25 can then be simplified by using the following equations:

$$\frac{1}{\alpha(w_k) \cdot \beta(w_k)} \cdot \left(\frac{\partial \alpha(w_k)}{\partial w_k} \cdot \beta(w_k) - \alpha(w_k) \cdot \frac{\partial \beta(w_k)}{\partial w_k} \right) =$$

$$= \frac{1}{\alpha(w_k) \cdot \beta(w_k)} \cdot \left(\alpha(w_k) \cdot f_k(\mathbf{d}_{j_{\mathcal{C}w_l}}) \cdot \beta(w_k) - \alpha(w_k) \cdot \frac{\partial \beta(w_k)}{\partial w_k} \right)$$

$$= f_k(\mathbf{d}_{j_{\mathcal{C}w_l}}) - \frac{1}{\beta(w_k)} \cdot \frac{\partial \beta(w_k)}{\partial w_k} \tag{3.32}$$

$$= f_k(\mathbf{d}_{j_{\mathcal{C}w_l}}) - \frac{\sum_{i^*=1}^{|\mathcal{D}_{X_i}|} \exp\left(\sum_{w_l \in \mathcal{W}_{X_i}} w_l f_l(\mathbf{d}_{j_{\mathcal{C}w_l}}, x_{i^*}) \right) \cdot f_k(\mathbf{d}_{j_{\mathcal{C}w_l}}, x_{i^*})}{\sum_{i^*=1}^{|\mathcal{D}_{X_i}|} \exp\left(\sum_{w_l \in \mathcal{W}_{X_i}} w_l f_l(\mathbf{d}_{j_{\mathcal{C}w_l}}, x_{i^*}) \right)}$$

Depending on the state represented by feature function f_k, the result of derivation 3.32 can then take one of three possible values:

1. If $f_k(\mathbf{d}_{j_{\mathcal{C}w_l}}) = 1$, then:

$$f_k(\mathbf{d}_{j_{\mathcal{C}w_l}}) - \frac{\sum_{i^*=1}^{|\mathcal{D}_{X_i}|} \exp\left(\sum_{w_l \in \mathcal{W}_{X_i}} w_l f_l(\mathbf{d}_{j_{\mathcal{C}w_l}}, x_{i^*}) \right) \cdot f_k(\mathbf{d}_{j_{\mathcal{C}w_l}}, x_{i^*})}{\sum_{i^*=1}^{|\mathcal{D}_{X_i}|} \exp\left(\sum_{w_l \in \mathcal{W}_{X_i}} w_l f_l(\mathbf{d}_{j_{\mathcal{C}w_l}}, x_{i^*}) \right)} =$$

$$= 1 - \frac{\exp\left(\sum_{w_l \in \mathcal{W}_{X_i}} w_l f_l(\mathbf{d}_{j_{\mathcal{C}w_l}}) \right)}{\sum_{i^*=1}^{|\mathcal{D}_{X_i}|} \exp\left(\sum_{w_l \in \mathcal{W}_{X_i}} w_l f_l(\mathbf{d}_{j_{\mathcal{C}w_l}}, x_{i^*}) \right)} \tag{3.33}$$

$$= 1 - P\left(x_{i_j} \mid mb_j(X_i) \right)$$

2. If $f_k(\mathbf{d}_{j_{\mathcal{C}w_l}}) = 0$ and $f_k(\mathbf{d}_{j_{\mathcal{C}w_l}}, x_i^*) = 1$ for some value x_i^* of X_i, then:

$$f_k(\mathbf{d}_{j_{\mathcal{C}w_l}}) - \frac{\sum_{i^*=1}^{|\mathcal{D}_{X_i}|} \exp\left(\sum_{w_l \in \mathcal{W}_{X_i}} w_l f_l(\mathbf{d}_{j_{\mathcal{C}w_l}}, x_{i^*}) \right) \cdot f_k(\mathbf{d}_{j_{\mathcal{C}w_l}}, x_{i^*})}{\sum_{i^*=1}^{|\mathcal{D}_{X_i}|} \exp\left(\sum_{w_l \in \mathcal{W}_{X_i}} w_l f_l(\mathbf{d}_{j_{\mathcal{C}w_l}}, x_{i^*}) \right)} =$$

$$= -\frac{\exp\left(\sum_{w_l \in \mathcal{W}_{X_i}} w_l f_l(\mathbf{d}_{j_{\mathcal{C}w_l}}, x_i^*) \right)}{\sum_{i^*=1}^{|\mathcal{D}_{X_i}|} \exp\left(\sum_{w_l \in \mathcal{W}_{X_i}} w_l f_l(\mathbf{d}_{j_{\mathcal{C}w_l}}, x_{i^*}) \right)} \tag{3.34}$$

$$= -P\left(x_i^* \mid mb_j(X_i) \right)$$

3. If $f_k(\mathbf{d}_{j_{\mathcal{C}w_l}}) = 0$ and $f_k(\mathbf{d}_{j_{\mathcal{C}w_l}}, x_{i^*}) = 0$ for any value x_{i^*} of X_i, then:

$$f_k(\mathbf{d}_{j_{\mathcal{C}w_l}}) - \frac{\sum_{i^*=1}^{|\mathcal{D}_{X_i}|} \exp\left(\sum_{w_l \in \mathcal{W}_{X_i}} w_l f_l(\mathbf{d}_{j_{\mathcal{C}w_l}}, x_{i^*}) \right) \cdot f_k(\mathbf{d}_{j_{\mathcal{C}w_l}}, x_{i^*})}{\sum_{i^*=1}^{|\mathcal{D}_{X_i}|} \exp\left(\sum_{w_l \in \mathcal{W}_{X_i}} w_l f_l(\mathbf{d}_{j_{\mathcal{C}w_l}}, x_{i^*}) \right)} = 0 \tag{3.35}$$

3.3.2 Finding the Maximal Cliques

It should be clear from the previous sections that a crucial component of Markov random fields is given by the maximal cliques of the graph. In order to enumerate the feature functions in a Markov random field, all maximal cliques of the graph must have been found. Unfortunately, the problem of finding all maximal cliques is NP-hard. In fact, accomplishing this task involves determining whether a graph contains at least a clique of a specified size s, which is one of the 21 problems that Richard Karp showed to be NP-complete [162].

A classic algorithm for finding maximal cliques in undirected graphs, developed by Coen Bron and Joep Kerbosch [37], is illustrated in Appendix B.2. Here, we present a particularly simple clique-finding strategy (see Algorithm 3.1). The proposed algorithm employs a bottom-up strategy to discover (maximal) cliques in an undirected graph. Starting from cliques of size 1, the idea is to merge all possible pairs of cliques of size s, checking whether their union forms a new clique (lines 6–11). On the other hand, for each s the algorithm discards those cliques of size $s - 1$ that are not maximal (lines 12–15). This algorithm may be a convenient choice when the ease of implementation is a major requirement. Moreover, empirical evidence suggests that its computational burden on a variety of graphs is not significantly higher than other state-of-the-art techniques.

Algorithm 3.1 findMaximalCliques: Finding all maximal cliques in undirected graphs

---

Input: Undirected graph $\mathcal{G}$ with nodes $X_1, \ldots, X_n$.
Output: Set containing all the maximal cliques of $\mathcal{G}$.

```
findMaximalCliques(G):
 1.    for(s = 1 to n)
 2.        γs = ∅
 3.    for(i = 1 to n)
 4.        γ1 = γ1 ∪ {{Xi}}
 5.    for(s = 1 to n-1)
 6.        for(Ci ∈ γs)
 7.            for(Cj ∈ γs \ {Ci})
 8.                C = Ci ∪ Cj
 9.                if(C is a clique)
10.                    γ|C| = γ|C| ∪ {C}
11.        for(Ci ∈ γs)
12.            for(Cj ∈ γs+1)
13.                if(Ci ⊂ Cj)
14.                    γs = γs \ {Ci}
15.    return ∪ⁿₛ₌₁ γs
```

3.4 Structure Learning

In Markov random fields, structure learning is the task of estimating the structure of the undirected graph from a given dataset. Since in Markov random fields parameter learning is much more complex than in Bayesian networks, heuristic structure learning of the kind we employ for directed models is computationally even more challenging in the case of undirected models. In fact, parameter learning involves optimizing a non-trivial objective function with respect to a possibly large number of (continuous-valued) parameters, and iterative structure learning clearly involves parameter learning as a subroutine. Given this, the usual way of fixing the structure of Markov random fields in their applications is by exploiting domain-specific knowledge in order to design an undirected graph that best fits the constraints of the domain at hand. For example, an overview of common strategies for applying Markov random fields to image processing is provided in [240]. The way we fix the graph structure for the purposes of our experimental demonstrations will be described in Chapter 6 (see Section 6.3.1). Nevertheless, an approximate approach to structure learning can be clearly defined (and sometimes even applied in practical situations) for Markov random fields in a similar way as we did in Section 2.4 for Bayesian networks. We now present an approach to this problem based on heuristic search in a space of possible graphs, as endorsed for example in [137]. To this aim, what we need to describe is: (*i*) the problem space, i.e. the set of graphs explored during the search; (*ii*) the heuristic function, i.e. the metric used for scoring different graphs; (*iii*) the search strategy, i.e. the method used for traversing the search space.

If we are dealing with a random vector $\mathbf{X} = (X_1, \ldots, X_n)$, then the search space for learning the structure of a Markov random field is nothing but the set of all possible (undirected) graphs wit h vertices $X_1, \ldots, X_n$. Within this set, we search for the graph that maximizes a specified evaluation function.

A simple choice for the evaluation function is the model pseudo-likelihood given the dataset, formulated in Equation 3.9. In order to prevent overfitting, we refine the evaluation function by introducing a regularization term, following the approach originally proposed by Reimar Hofmann and Volker Tresp for continuous Markov networks [137]. We then come up with the following measure, which we denote by MDL^* (since it is inspired by the same principles leading to the MDL/BIC score, as discussed in Section 2.4.1):

$$MDL^*(h) = \log P^*(\mathbf{D} \mid h) - \lambda \frac{DL(h)}{2} \log |\mathbf{D}| \qquad (3.36)$$

where h is a Markov random field to be scored, $DL(h)$ is the model description length, as measured by the number of edges in h, and λ is a regularization parameter (to be tuned by suitable cross-validation trials). A convenient feature of the suggested scoring function is that it does not require any explicit assessment of the model prior probability.

In order to search the model space, we employ a greedy hill-climbing technique, based on tentatively removing edges from an initial, fully connected graph.

When we remove an edge linking two nodes X_i and X_j, both the structure of $\mathcal{MB}(X_i)$ and $\mathcal{MB}(X_j)$ are affected. Therefore, in order to verify whether the edge removal improves the model score, we simply need to recompute the values of $\sum_{k=1}^{|\mathbf{D}|} \log P\left(x_{i_k} \mid mb_k(X_i)\right)$ and $\sum_{k=1}^{|\mathbf{D}|} \log P\left(x_{j_k} \mid mb_k(X_j)\right)$ respectively. Additionally, if we use the MDL-based metric given in Equation 3.36, we also need to recompute the regularization term based on the new value of $DL(h)$.

The greedy search proceeds as follows. Given a current Markov network h, we construct the set $\mathcal{H}$ containing all Markov networks resulting from removing exactly one edge from h. We then compute the score $MDL^*(h_i)$ for each h_i contained in $\mathcal{H}$, and denote by h^* the highest-scoring element of $\mathcal{H}$. If h^* scores better then h, we replace h by h^* and repeat the described cycle, otherwise the algorithm stops and h is returned as output. Pseudocode for the described technique is provided by Algorithm 3.2, where the notation $\mathcal{E}_h$ refers to the set of edges contained in Markov random field h. Note that the presented algorithm differs from the one proposed in [137] in the following respect. Suppose that h^* is such that $\mathcal{E}_{h^*} = \mathcal{E}_h \setminus \{\{X_i, X_j\}\}$. Then, the latter algorithm only replaces h by h^* if both $\sum_{k=1}^{|\mathbf{D}|} \log P\left(x_{i_k} \mid mb_k(X_i)\right)$ and $\sum_{k=1}^{|\mathbf{D}|} \log P\left(x_{j_k} \mid mb_k(X_j)\right)$ increase after removing the edge $\{X_i, X_j\}$, which is a stronger requirement than the one imposed by Algorithm 3.2.

Algorithm 3.2 `HCLearnMRF`: Hill-climbing structure learning in Markov random fields

---

Input: Dataset $\mathbf{D}$; Markov random field h with a fully connected graph.
Output: Markov random field h maximizing the heuristic function $MDL^*(h)$ with respect to $\mathbf{D}$.

```
HCLearnMRF(D, h):
1.   do
2.       s = MDL*(h)
3.       H = {h_i : E_{h_i} ⊂ E_h ∧ |E_{h_i}| = |E_h| - 1}
4.       h* = arg max_{h_i ∈ H} MDL*(h_i)
5.       s* = MDL*(h*)
6.       if (s* > s)
7.           h = h*
8.   while (s* > s)
9.   return h
```

3.5 Final Remarks

In summary, Markov random fields are graphical models that, analogously to Bayesian networks, are suitable to represent joint distributions over sets of random variables. While BNs rely on directed, acyclic graphical structures, the graph underlying Markov networks is undirected in nature. Also, while the edges in BNs tend to express a sort of semantic relationship between pairs of variables (e.g., a causal link), in MRFs the relation is symmetric and mostly related to the topological properties of neighboring variables over a space (that is, the mathematical notion of a random 'field'). Cyclic dependencies cannot be captured in BNs, but they actually can in MRFs. In other words, there are structures of (in)dependencies over a set of variables that can be represented by means of Markov networks, but that BNs cannot represent. This observation could lead to the suspect that BNs are just a special case of MRFs. In the next chapter we will see that this is not the case. This will provide us with further motivation for developing a novel graphical model which can suitably represent both the families of independence properties representable by BNs and MRFs.

We saw that the class of Markov random fields considered in this book express a joint distribution as a normalized product (extended to the maximal cliques in the graph) of clique-specific potential functions ('local energies'), having the form of exponential functions[5]. These exponentials are, in turn, defined over linear combinations of the feature functions, acting as boolean indicators of the variables in a clique being in a specific state. The quantity used for the normalization of the product expressing the joint density is the partition function. The reader is reminded of the qualitative introduction to the concept of MRF we made in Section 3.1, since these formal definitions are instances of the corresponding notions rooted in statistical mechanics. In particular, as anticipated, Equation 3.7 is a Gibbs distribution having the same form as Equation 3.2. In going through Section 3.1, the reader might wonder whether a precise relation holds between the two qualitative examples we worked out in order to introduce the concept of MRF, i.e. the distribution of temperatures at different locations on Earth (relying on the assumption of certain Markovian properties) and the statistical mechanics models (whose probability is modeled by assuming a Gibbs distribution). We saw that the Hammersley-Clifford theorem provides us with the mathematical proof of equivalence between the two perspectives and the corresponding statistical assumptions, once the probability density function is strictly positive.

We also saw that, in general, computing the partition function (hence, the joint distribution) is intractable, except for toy situations which are of limited practical relevance. This problem was tackled after Besag's guidelines, i.e. resorting to a much simpler, yet well-behaved pseudo-likelihood measure. The pseudo-likelihood relies on the calculation of the conditional probability of each variable given its Markov blanket (a notion we already met in BNs). A parameter learning algorithm is then needed for MRFs. The parameters to be learnt from the data are the real-valued weights associated with the feature functions. In this perspective, Markov networks define a parametric model with parameters $w_1, \ldots, w_l$. We then adopted a maximum-(pseudo-)likelihood optimization criterion, and we applied gradient-based methods for estimating the 'optimal' weights. In particular, we saw that the limited memory BFGS algorithm may be a reasonable choice (it exhibits limited memory requirements, allowing also for large-scale applications). To this end, we

---

[5] Note that the factorization of the joint distribution over the cliques is guaranteed by the specific definition of MRFs we adopted in the book, sometimes referred to as the log-linear model. In general, not all MRFs can be factorized accordingly.

outlined explicit calculations for the partial derivatives of the criterion function with respect to the weights of the model. The results are not limited to L-BFGS; in fact, they can be applied to any gradient-based optimization scheme. In order to come up with a proper definition of the potential functions in terms of specific weights (and, in order to make parameter learning a meaningful task), maximal cliques in the graph must be known or discovered. Unless prior knowledge of the application domain helps pinpointing them, a technique for finding the maximal cliques is needed. We introduced a very simple, yet effective algorithm for accomplishing this task.

As in Bayesian networks, structure learning is another critical issue in MRFs. In the literature, a specific undirected graphical structure is mostly assumed, relying on prior knowledge. Nonetheless, in the general case a structure learning algorithm is required. Tho this end, the technique we developed is inherited from the approach we took in the previous chapter for learning the structure of BNs. The algorithm realizes a search strategy in a space of undirected graphs. The search should be guided by the (pseudo-)likelihood heuristic. Unfortunately, from a computational point of view, the problem is even harder than it used to be in BNs. This is due to (*i*) the nature of the potential functions, and (*ii*) the need for continuous-valued weights, obtained via gradient-based optimization. For these reasons, we proposed an approximate algorithm, aimed at maximizing a regularized version of the pseudo-likelihood scoring metric.

3.5.1 Generalizations and Variations on the Theme

Like in the case of BNs, the vast majority of the (machine learning) literature on Markov networks is concerned with discrete random variables. Nevertheless, MRFs can be generalized to continuous-valued variables, without any significant alterations of the overall nature and mathematical properties of the model. We deal with continuous-valued graphical models in Chapter 5.

A number of generalizations to standard Markov networks have been proposed in the literature. In the following, we limit our attention to some of the most representative variations on the theme of MRFs. Multiscale representation of MRFs is discussed in [199], where scale-recursive algorithms are introduced that allow for a unified treatment of higher-dimensional Markov networks in an efficient manner. Discriminative random fields for pattern classification are introduced in [183]. They rely on local discriminative models, and they relax the usual assumption of mutual independence among the data given the class that is typical in the application of Markov networks to pattern recognition problems. MRFs and support vector machines are combined within a unifying framework, called 'M3' (Maximum margin Markov network), in [295]. M3 stems from the idea of exploiting the generalization properties of support vector machines and their discriminative power in classification tasks, along with the capability of MRFs of modeling statistical dependencies among the variables.

Hidden Markov random fields (HMRF) [185] can be seen as either an extension of MRFs, or as a generalization of hidden Markov models. Bearing in mind the definition of HMM we gave in Section 2.6.1, the idea can be easily formulated as follows. A hidden Markov random field is defined exactly as a hidden Markov model whose hidden part (the Markov chain) is generalized to be a Markov random field, i.e. a lattice of undirected statistical dependencies over discrete, latent state variables. Like in HMMs, each state variable has its own emission probability distribution (which is independent of the other variables given the state itself). The set of the emission probabilities forms the observable part of the model. Of course, both training and decoding algorithms are adapted to the new, extended framework [185]. Model selection criteria for HMRFs, based on mean field-like approximations, are presented in [91].

Hierarchical extensions of Markov random fields were recently introduced [327, 274]. In particular, dynamic hierarchical Markov random fields allow accounting for structural uncertainty and modeling the structure and the class-labels according to a joint distribution. The approach was successfully evaluated in a web-based data extraction task [327].

Gaussian Markov random fields [24, 55, 27, 266, 7] are MRFs such that the joint probability distribution of the variables can be modeled explicitly with a multivariate Normal distribution. (Relevance of Normal distributions to continuous-valued variables in graphical models is dramatic, and the topic will be covered extensively in Section 5.3.1). The equivalence between a Markov network and a multivariate Gaussian can be shown to hold true whenever the missing edges in the graph underlying the MRF coincide with the zeros of the inverse of the covariance matrix of the corresponding Normal density.

A particularly intriguing generalization of the Markov random field model is the *conditional random field* (CRF), which has become a popular alternative to hidden Markov models for sequence recognition and labeling (especially in natural language processing and bioinformatics applications) [186, 294]. The simplest way to describe a CRF is by defining it as an extension of a traditional Markov network where the probability distribution is a *conditional density* (instead of being a bare joint density over the nodes of the graph) given the value of 'external' random variables (i.e., observations of the phenomenon under consideration, like the acoustic observations in a HMM). Each potential function in the conditional random field is, in turn, defined over the variables of its clique *and* the external variables. The graphical structure of CRFs may be arbitrary, just like in ordinary MRFs, but in most practical applications it reflects the sequential nature of the data (a discrete-time random process), assuming the form of an undirected chain. While individual nodes in the topology of HMMs do not have, in general, any special meaning (e.g., a state node in a left-to-right HMM for speech recognition takes responsibility for just a short, unlabeled portion of a sub-phonetic unit), in CRFs each vertex is usually interpreted as a well-defined tag, or label. Conditional random fields are discriminative models [186, 294], while standard HMMs and MRFs are not. This is usually claimed to be the reason why CRFs are better suited to classification/tagging/labeling tasks over sequences [186, 294]. By assuming a chain structure of dependencies between pairs of adjacent variables, efficient algorithms exist for solving the CRF-equivalent of the well-known 'three fundamental problems' of HMMs, formulated by Rabiner in 1989 [251]: (1) computing the probability of an input observation sequence given the model (i.e. *inference*); (2) finding the optimal sequence of hidden variables that best explains the input sequence (i.e. *decoding*); (3) estimating the model parameters that best fit the observation sequences of the training set, according to the maximum-likelihood criterion (i.e. *learning*). Actually, from a theoretical point of view HMMs can be seen as a particular case of CRFs, while the latter basically extend the former by allowing for: (*i*) transition probabilities that are not fixed, since they can change 'in time' (i.e., at different positions along the input sequence); (*ii*) potential functions (that replace the emission densities of HMMs) that can take into account input observations at any depth back in time, and that are not constrained to satisfy the axioms of probability (since they are not forced to have a straightforward probabilistic interpretation). In practice, there are limitations to the application of CRFs in the place of HMMs, in particular when no semantic interpretation of individual observations can be given (e.g., speech processing, handwriting recognition), and/or when it is not clear (from the knowledge of the application domain) how many observations back in time shall the feature functions take into account. Furthermore, like in MRFs, the learning problem turns out to be complex, and its approximate, gradient-based solutions may result to be (sometimes, way too) suboptimal. Efficient techniques for inducing the features in CRFs are investigated in [209]. Conditional random fields may be combined with support vector machines too, taking benefit from the generalization capabilities of the latter (see, e.g., [193]).

Other interesting, implicit (and often overlooked) variations on the theme of Markov networks can be found in the area of artificial neural networks (ANNs). A thorough treatment of ANNs is beyond the scope of this book. Nonetheless, it is worth making a few, major concepts clear. Let us say that an artificial neural network can be described as a graph [258]. According to the specific family of ANNs we consider, the graphical structure may be either directed (e.g., multilayer perceptrons [268]) or undirected (e.g., Boltzmann machines [1]), as well as cyclic (e.g., recurrent ANNs [239, 79]) or acyclic (e.g., feedforward ANNs [29]). Each vertex in the graph (called neuron, or unit) represents the state of an input random variable (input units), or a random quantity (similarly to what happens in Bayesian or Markov networks with latent variables), affected by the states of other units (hidden, or output neurons). The graph is labeled, meaning that a real-valued label ('weight', or 'connection strength') w_{ij} is associated with each edge ('synaptic connection') between j-th and i-th neurons. The weight expresses the amount of 'correlation' between the corresponding units, and whether the correlation is positive or negative. An activation (or potential) function $f(.)$ is associated with each node in the graph. The argument x_i of $f_i(.)$ for the generic i-th unit is, in most cases, a weighted linear combination of the activation of 'neighboring' nodes, namely $\sum_j w_{ij} f_j(x_j)$, where the sum is extended to all nodes j in the graph such that (j, i) is in the set of edges. The argument of the activation function reminds us of the corresponding argument of potential functions in MRFs. Although a variety of activation functions have been proposed in the literature, conferring different nature and properties to the ANNs, the most popular potential functions are the sigmoid, i.e. $f(x) = \frac{1}{1+\exp(-x)}$, and the Gaussian (i.e., an exponential). From the perspective of this book, it is crucial to observe that a probabilistic interpretation can be readily given to both of them. In point of fact, the Gaussian *is* the probability density function of its input, while the sigmoid is the cumulative distribution function of the corresponding logistic density $\frac{\exp(-x)}{(1+\exp(-x))^2}$ (that is, the sigmoid implicitly entails a specific probability distribution of its inputs). ANNs are trained from data, meaning that the value of their weights is learned (usually applying the gradient method) in order to extremize a certain global energy function (a.k.a. *criterion*, or *error function*) which is affected by the local activations.

ANNs may be used as an (explicit or implicit) model of the probability density of the data. Major instances are: the probabilistic neural network [284], which is a connectionist representation of the Parzen window estimate [73]; radial basis function networks [220], which realize mixtures of Gaussian densities, and that can be trained over the maximum-likelihood criterion in order to estimate probability densities [303]; and, multilayer perceptrons having a probabilistic interpretation as non-parametric models of Bayesian posterior probabilities [255, 31], or of density functions [302, 297]. In these cases, it is possible to say that ANNs are probabilistic graphical models that can represent joint probability distributions over a set of random variables (namely, the components of the input vectors). It is noteworthy that, in general, they do not make any prior independence assumptions on the input variables. Out of the wide class of graphical models that can be realized via ANNs, we can spot at least two important families of neural nets which turn out to be special cases of Markov random fields, namely Hopfield networks, and Boltzmann machines.

The Hopfield network [141, 84] is a flat lattice of interconnected, deterministic units. Each unit (acting as an input *and* an output neuron at a given 'location') receives a single scalar input (namely, a binary value) and returns a $-1/+1$ output, yielded by the application of a threshold activation function to the weighted sum of contributions from the other units. The output affects all the other units according to a symmetric topology of undirected edges, which form a cyclic graph (that is, the network is recurrent). A Lyapunov (i.e., energy) function is associated with each possible state of the network, whose probability can be shown to follow the Gibbs distribution (i.e., the network possesses the Markov properties). When fed with a new input pattern, the network behaves like a dynamical system and tends to stabilize around an equilibrium point (either a basin of attraction, or a spurious state) of the energy function. It is an instance of the Ising model we introduced in Section 3.1 and, in turn, an instance of MRF. Training the network from data means developing connection weights that allow the model to reach minima of the energy function, so that the machine memorizes the input patterns. A given Hopfield architecture can memorize a limited amount of different patterns, turning out to have a maximum memory capacity [133]. An input stimulus that partially matches a memorized pattern will cause the model to recover the most similar pattern in the memory of the network. This phenomenon induces an autoassociative memory behavior, which can reconstruct complete patterns from partial hints. For this reason, this particular graphical model has often been referred to as a content addressable memory.

While Hopfield nets have deterministic activation functions, nondeterministic potential functions (whose random behavior is governed by the Boltzmann-Gibbs distribution) are found in another ANN instance of the MRF, namely the Boltzmann machine [1, 133]. The graph underlying Boltzmann machines has basically the same structure and meaning it has in Hopfield networks. The differences arise at the levels of potential functions and global energy. The state of each unit is actually a random variable, drawn from the Boltzmann-Gibbs distribution as a function of a certain temperature of the system. The general idea is that the network behavior is the consequence of a simulated annealing process, where the machine is run first under high temperature (entailing a strong stochastic behavior), and progressively cooled-down to a thermal equilibrium. Each unit is not limited to binary values, but can take any real values. Like the Markov networks we discussed in the previous Sections, training of Boltzmann machines is aimed at maximizing the (log-)likelihood criterion, practically exploiting a Gibbs sampling procedure.

3.5.2 Applications of Markov Random Fields

Markov networks found a number of applications. It is easily seen that most of them concern domains where the topology of the MRF fits a spatial representation of the data in a natural way. In this respect, the most relevant instances are rooted in image processing (e.g., [106]), although we should not underestimate the fact that, as we say, several paradigms suitable for sequence processing (dynamic BNs, HMMs) can be thought of as special cases of MRFs. They were intensively applied to such areas as speech recognition [119], natural language processing [156], handwriting recognition [325], and bioinformatics [12]. In the following we report a few, major examples.

A tutorial on using MRFs for stochastic image modeling can be found in [240]. In [66], MRFs are applied to range sensing, in order to reconstruct high resolution range images, having low noise, from the integration of low resolution, noisy range image captured via ordinary cameras. A MRF is used for image segmentation and skin detection as part of a system for human emotion recognition from color images in [204]. An application to image processing of multi-dimensional multivariate Gaussian Markov random fields is presented in [205]. Segmentation of images relying on Bayesian smoothing modeled by MRFs is discussed in [64]. Recently, MRFs were applied to an image retrieval task involving also global image features [198]. Segmentation of brain MR images via Markov networks is investigated in [130]. Self-validated labeling of Markov random fields is proposed and applied to image segmentation in [88]. Again, a tree-structured MRF model is developed in [61], where it is used for Bayesian image segmentation. Application to pixon-based segmentation of images is treated in [320]. Another segmentation model, this time based on compound Markov random fields and a boundary model, can be found in [319].

A number of applications of MRFs, and of related models, to semi-supervised learning problems are surveyed in [328]. Furthermore, MRFs were applied to such exciting areas as sonar signal analysis for the classification of underwater floor [215], landmine detection in ground penetrating radar data [296], fast damage mapping in case of earthquakes [304], and automatic detection of pornographic images [324].

A generative model for separating illumination and reflectance from images is proposed in [288], where hidden Gaussian Markov random fields are combined with a generalized autoregressive process. Integrated Web data extraction via dynamic hierarchical MRFs is accomplished in [327]. Adaptive Gaussian Markov random fields are presented and successfully applied to human brain mapping in [34].

3.5.3 Points of Strength and Limitations of Markov Random Fields

Markov random fields are, along with Bayesian networks, among the most popular and effective graphical models. Aside from their theoretical properties, the variety of applications listed in the previous section makes a clear point on their practical relevance to a wide range of real-world challenges. Their application is particularly well suited, as we say, to domains like image processing, where the overall probability of the input pattern is defined in terms of local quantities which depend on the location over a topological space. In most cases, the very nature of the application domain is reflected in a specific, well-defined choice for the graphical structure. We saw that Markov networks subsume several instances of BNs and dynamic BNs, as well. While BNs are limited to acyclic graphs, MRFs can cope with cyclic dependence relationships in a natural way. Several generalizations to the definition of MRFs are presented in the literature, which further extend the potential of the model. Several popular learning machines, including HMMs and some families of neural networks, can be seen as special cases of MRFs, too. Adoption of Besag's pseudo-likelihood metric, along with suitable algorithms for finding maximal cliques and searching the space of graph structures, confers flexibility to the model, which can be virtually applied to severe tasks even in the absence of prior knowledge.

Nevertheless, just like in the case of BNs, there are several drawbacks that may limit the application of Markov networks to some (possibly significant) extent. First of all, Markov random fields are usually designed (as we did throughout the chapter) for discrete-valued variables. Extensions to the continuous-valued case (see Chapter 5) do not affect the basic definitions and the main algorithms, but they require some non-trivial workarounds. In fact, there is no standard way to realize the extension (especially, as far as the modeling of real-valued probabilistic quantities is concerned), and the overall complexity of learning turns out to be increased further.

Complexity, indeed, is an issue in MRFs. Its darkest side is the computation of the partition function. Resorting to the pseudo-likelihood tackles the problem only partially, and is doomed to lead to approximate, sub-optimal solutions in practice. Parameter and structure learning are also complex problems, and their algorithmic solutions do not scale-up well with the number of variables in the graph. For instance, no general solution to this problem has ever been found for Boltzmann machines, forcing the community to limit applications to a sub-class of these models, known as the restricted Boltzmann machine. Information-theoretic limits on MRF structure learning over high dimensions are derived in [270]. In the next chapter, scalability of learning to high-dimensional spaces is going to be one of our primary concerns in designing the hybrid random field model.

Another relevant limitation of MRFs is that, in spite of the fact that several BNs can be represented within the present framework, there are independence relationships that can be modeled properly via Bayesian networks, but that do not fit the MRF modeling capabilities. Instances may be a consequence of the undirected nature of the edges of the graph underlying the definition of MRF. Undirected edges do not allow for a representation of causal relations, limiting the expressiveness of MRFs (at least from a conceptual point of view), e.g. when an expert's knowledge (on causal effects of certain variables on others) shall be encapsulated within an expert system capable of probabilistic reasoning. In a broader sense, while BNs tend to capture the semantics underlying the graphical representation of the problem domain by means of specific variables and their 'ordered' relationships, MRFs tend to be silent concerning a possible ordering over the relationships. Again, hybrid random fields will cover all possible independence relations that can be modeled with BNs or MRFs altogether. It goes without saying that MRFs can model joint probability distributions over a set of variables only under the Markov assumption, and in particular (due to the Hammersley-Clifford theorem) only when the involved distribution is strictly positive.

Finally, Markov random fields are intrinsically non-discriminative, which may limit their effective application to pattern classification tasks. This is possibly the utmost rationale behind the recent development of discriminative generalizations of Markov networks, such as discriminative random fields [183], max-margin Markov networks [295], and conditional random fields [186], capable of modeling conditional (i.e, posterior) probabilities—instead of bare joint densities—in an explicit manner.

Chapter 4
Introducing Hybrid Random Fields: Discrete-Valued Variables

> *"Es gibt keine tabula rasa. Wie Schiffer sind wir, die ihr Schiff*
> *auf offener See umbauen müssen, ohne es jemals in einem Dock*
> *zerlegen und aus besten Bestandteilen neu errichten zu können."*
>
> Otto Neurath, 1932 [227]

4.1 Introduction

Both Bayesian networks and Markov random fields are widely used tools for statistical data analysis. Applying Markov random fields can be relatively expensive because of the computational cost of learning the model weights from data. As we saw in Chapter 3, this task involves optimization over a set of continuous parameters, and the number of these parameters can become very large (in the order of several thousands and more) when we try to address problems involving several variables. For example, the link-prediction application we will describe in Section 6.3.4.3, which involves 1,682 variables, is such that the Markov random field applied to it contains 16,396 feature functions, with the corresponding weights.

Two clear advantages of Bayesian networks over Markov random fields are on the one hand the relative ease of parameter learning, and on the other hand (based on the former advantage) the feasibility of learning the model structure using heuristic algorithms (at least when dealing with datasets of reasonable dimensionality). However, a major limitation of Bayesian networks is the significant computational cost of structure learning in high-dimensional domains. In order to gain insight into the mathematical nature of this problem, it is useful to consider a result in graph theory, established in [260]. The result states that the total number of DAGs containing n nodes is given by a recursive function $f(n)$, defined as follows:

$$f(n) = \begin{cases} 1 & \text{if } n = 0 \\ \sum_{i=1}^{n} -1^{i+1} \binom{n}{i} 2^{i(n-i)} f(n-i) & \text{if } n > 0 \end{cases} \quad (4.1)$$

This means that the size of the search space for the structure learning algorithm described in Section 2.4 grows exponentially with the number of nodes in the DAG [50]. To get an idea of how large that space becomes as n increases, notice for example that $f(3) = 25$, $f(5) = 29,000$, and $f(10) = 4.2 \cdot 10^{18}$ [226]. As a consequence, although Algorithm 2.1 is relatively fast due to the hill-climbing strategy, we cannot

A. Freno and E. Trentin: Hybrid Random Fields, ISRL 15, pp. 69–86.
springerlink.com   © Springer-Verlag Berlin Heidelberg 2011

expect it to scale well to high-dimensional domains. Such limitation will be apparent from the experimental analysis provided in Chapter 6.

This scenario provides the main motivation for turning to the hybrid graphical model we describe in this chapter. That is, the model we are going to present is aimed at overcoming the limitations of both Bayesian networks and Markov random fields in terms of the computational cost of estimating the model from data. To this aim, the basic strategy behind hybrid random field estimation is to exploit on the one hand the ease of learning Bayesian networks for relatively small domains, and on the other hand the factorization of joint distributions employed in Markov random fields in the form of the pseudo-likelihood measure. In particular, hybrid random fields model the conditional distribution of each variable X_i given its Markov blanket in $\mathbf{X}$ through a *local* Bayesian network, i.e. a Bayesian network containing only a suitable subset of variables from $\mathbf{X}$.

The chapter is structured as follows. Section 4.2 presents the mathematical definition of hybrid random field, after introducing the preliminary concepts of directed and undirected union over graphs. Then, a fundamental theorem is proved, to the effect that (*i*) HRFs provide a well-defined representation of joint probability distributions, and (*ii*) the set of conditional independence statements entailed by any given HRF is graphically identified by a statistical property called 'modularity condition'. In other words, we show that any HRF can be mapped onto a unique joint distribution, characterized by an explicit set of conditional independencies underlying the modeled variables, and we explain how the represented distributions can be derived from the HRF. Section 4.3 offers a mathematical analysis of the main theoretical properties satisfied by the (conditional) independence structures modeled by hybrid random fields. The ultimate contribution of the analysis lies in showing that hybrid random fields are capable of modeling exactly the union of the classes of independence structures that can be modeled by Bayesian and Markov networks respectively. This result may be summarized by saying that the modularity condition is mathematically equivalent to the logical disjunction of the two statistical properties that we referred to as (*i*) directed Markov assumption (see Section 2.2) and (*ii*) local/global Markov property (see Section 3.2.1). After completing the formal analysis of the defined graphical model, we proceed to presenting suitable inference and learning algorithms for hybrid random fields. Concerning inference, we describe the ordered Gibbs sampling technique (Section 4.4), whose generality allows to apply it to hybrid random fields in exactly the same way it is traditionally applied to other graphical models, such as Bayesian and Markov networks. Given the directed-graphical nature of the local modules composing HRFs, we show how parameter learning can be addressed by simply exploiting the estimation technique we already presented for Bayesian networks. On the other hand, in order to estimate the structure of HRFs, Section 4.6 presents a dedicated algorithm, known as Markov Blanket Merging [95, 94], which is explained to exhibit nice properties in terms of computational complexity (Section 4.6.4). Section 4.7 briefly discusses—especially from a machine learning point of view—some analogies and differences between HRFs and a few other graphical models that have been proposed in the relevant literature, such as dependency networks and chain graphs. Finally, while summarizing the main results collected throughout the chapter, Section 4.8 also discusses some important differences between the formal definition and theoretical analysis of hybrid random fields offered in Sections 4.2–4.3 and the respective definition and analysis which can be found in the literature originally introducing the model to the scientific community [94, 97].

4.2 Representation of Probabilities

Just like Bayesian networks and Markov random fields, hybrid random fields are aimed at representing joint probability distributions underlying sets of random variables. In order to define the concept of hybrid random field, we need to introduce a few preliminary notions concerning directed and undirected graphs. First, let us define the concept of *(directed) union* of directed graphs:

Definition 4.1. If $\mathcal{G}_1 = (\mathcal{V}_1, \mathcal{E}_1)$ and $\mathcal{G}_2 = (\mathcal{V}_2, \mathcal{E}_2)$ are directed graphs, then the *(directed) union* of $\mathcal{G}_1$ and $\mathcal{G}_2$ (denoted by $\mathcal{G}_1 \cup \mathcal{G}_2$) is the directed graph $\mathcal{G} = (\mathcal{V}, \mathcal{E})$ such that $\mathcal{V} = \mathcal{V}_1 \cup \mathcal{V}_2$ and $\mathcal{E} = \mathcal{E}_1 \cup \mathcal{E}_2$.

Clearly, for a set of directed graphs $\mathcal{G}_1, \ldots, \mathcal{G}_n$, the directed union $\mathcal{G} = \bigcup_{i=1}^n \mathcal{G}_i$ simply results from iterated application of the binary union operator. Similarly, we define the notion of *(undirected) union* for undirected graphs:

Definition 4.2. If $\mathcal{G}_1 = (\mathcal{V}_1, \mathcal{E}_1)$ and $\mathcal{G}_2 = (\mathcal{V}_2, \mathcal{E}_2)$ are undirected graphs, then the *(undirected) union* of $\mathcal{G}_1$ and $\mathcal{G}_2$ (denoted by $\mathcal{G}_1 \cup \mathcal{G}_2$) is the undirected graph $\mathcal{G} = (\mathcal{V}, \mathcal{E})$ such that $\mathcal{V} = \mathcal{V}_1 \cup \mathcal{V}_2$ and $\mathcal{E} = \mathcal{E}_1 \cup \mathcal{E}_2$.

Now, if $\mathcal{G} = (\mathcal{V}, \mathcal{E})$ is a directed graph, then we say that $\mathcal{G}^* = (\mathcal{V}^*, \mathcal{E}^*)$ is the *undirected version* of $\mathcal{G}$ if $\mathcal{V}^* = \mathcal{V}$ and $\mathcal{E}^* = \{\{X_i, X_j\} : (X_i, X_j) \in \mathcal{E}\}$. Thus, we also define the undirected union for a pair of directed graphs:

Definition 4.3. If $\mathcal{G}_1 = (\mathcal{V}_1, \mathcal{E}_1)$ and $\mathcal{G}_2 = (\mathcal{V}_2, \mathcal{E}_2)$ are directed graphs, then the undirected union of $\mathcal{G}_1$ and $\mathcal{G}_2$ (denoted by $\mathcal{G}_1 \uplus \mathcal{G}_2$) is the (undirected) graph $\mathcal{G}^* = \mathcal{G}_1^* \cup \mathcal{G}_2^*$ such that $\mathcal{G}_1^*$ and $\mathcal{G}_2^*$ are the undirected versions of $\mathcal{G}_1$ and $\mathcal{G}_2$ respectively.

Given the concepts introduced above, a hybrid random field can be defined as follows:

Definition 4.4. Let $\mathbf{X}$ be a set of random variables $X_1, \ldots, X_n$. A *hybrid random field* for $X_1, \ldots, X_n$ is a set of Bayesian networks $BN_1, \ldots, BN_n$ (with DAGs $\mathcal{G}_1, \ldots, \mathcal{G}_n$) such that:

1. Each BN_i contains X_i plus a subset $\mathcal{R}(X_i)$ of $\mathbf{X} \setminus \{X_i\}$;
2. If $\mathcal{MB}_i(X_i)$ denotes the Markov blanket of X_i in BN_i (i.e. the set containing the parents, the children, and the parents of the children of X_i in $\mathcal{G}_i$) and $P(X_i \mid \mathcal{MB}_i(X_i))$ is the conditional distribution of X_i given $\mathcal{MB}_i(X_i)$, as derivable from BN_i, then at least one of the following conditions is satisfied:

a. The directed graph $\mathcal{G} = \bigcup_{i=1}^{n} \mathcal{G}_i$ is acyclic and there is a Bayesian network $h_\mathcal{G}$ with DAG $\mathcal{G}$ such that, for each X_i in $\mathbf{X}$, $P(X_i \mid \mathcal{MB}_i(X_i)) = P(X_i \mid \mathcal{MB}_\mathcal{G}(X_i))$, where $\mathcal{MB}_\mathcal{G}(X_i)$ is the Markov blanket of X_i in $h_\mathcal{G}$ and $P(X_i \mid \mathcal{MB}_\mathcal{G}(X_i))$ is the conditional distribution of X_i given $\mathcal{MB}_\mathcal{G}(X_i)$, as entailed by $h_\mathcal{G}$;

b. There is a Markov random field $h_{\mathcal{G}^*}$ with graph $\mathcal{G}^*$ such that $\mathcal{G}^* = \mathcal{G}_1 \uplus \ldots \uplus \mathcal{G}_n$ and, for each X_i in $\mathbf{X}$, $P(X_i \mid \mathcal{MB}_i(X_i)) = P(X_i \mid \mathcal{MB}_{\mathcal{G}^*}(X_i))$, where $\mathcal{MB}_{\mathcal{G}^*}(X_i)$ is the Markov blanket of X_i in $h_{\mathcal{G}^*}$ and $P(X_i \mid \mathcal{MB}_{\mathcal{G}^*}(X_i))$ is the conditional distribution of X_i given $\mathcal{MB}_{\mathcal{G}^*}(X_i)$, as derived from $h_{\mathcal{G}^*}$.

The elements of $\mathcal{R}(X_i)$ are called 'relatives of X_i'. That is, the relatives of a node X_i in a hybrid random field are the nodes appearing in graph $\mathcal{G}_i$ (except for X_i itself). An illustration of a hybrid random field is provided in Figure 4.1.

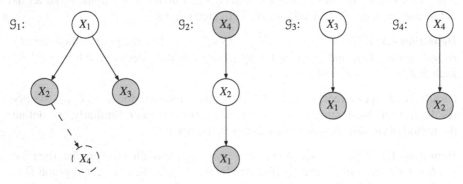

Fig. 4.1 The graphical components of a hybrid random field for the variables $X_1, \ldots, X_4$. Since each node X_i has its own Bayesian network (where nodes in $\mathcal{MB}_i(X_i)$ are shaded), there are four different DAGs. Relatives of X_i that are not in $\mathcal{MB}_i(X_i)$ are dashed.

The crucial issue concerning Definition 4.4 can be stated by means of two (tightly correlated) questions:

1. Is it possible to extract a joint probability distribution from a hybrid random field?
2. Is it possible to characterize a set of conditional independence statements entailed by a hybrid random field?

Both questions are given a positive answer by the following theorem:[1]

Theorem 4.1. *Suppose that h is a hybrid random field for the random vector $\mathbf{X} = (X_1, \ldots, X_n)$, and let h be made up by Bayesian networks $BN_1, \ldots, BN_n$ (with DAGs $\mathcal{G}_1, \ldots, \mathcal{G}_n$). Then, if each conditional distribution $P(X_i \mid \mathcal{MB}_i(X_i))$ is strictly positive (where $1 \le i \le n$), h has the following properties:*

---

[1] The theorem exploits the concept of an ordered Gibbs sampler. The reader who is not familiar with Gibbs sampling techniques will find it helpful to read Section 4.4 first.

1. *An ordered Gibbs sampler applied to h, i.e. an ordered Gibbs sampler which is applied to the conditional distributions $P(X_1 \mid \mathcal{MB}_1(X_1)), \ldots, P(X_n \mid \mathcal{MB}_n(X_n))$, defines a joint probability distribution over $\mathbf{X}$ via its (unique) stationary distribution $P(\mathbf{X})$;*
2. *For each variable X_i in $\mathbf{X}$, $P(\mathbf{X})$ is such that $P(X_i \mid \mathbf{X} \setminus \{X_i\}) = P(X_i \mid \mathcal{MB}_i(X_i))$.*

Proof. By the definition of hybrid random field, at least one of the following conditions must hold for any i such that $1 \leq i \leq n$:

1. $P(X_i \mid \mathcal{MB}_i(X_i)) = P(X_i \mid \mathcal{MB}_{\mathcal{G}}(X_i))$, where $\mathcal{G}$ is the directed union of $\mathcal{G}_1, \ldots, \mathcal{G}_n$ and $P(X_i \mid \mathcal{MB}_{\mathcal{G}}(X_i))$ is derived from a Bayesian network $h_{\mathcal{G}}$ with DAG $\mathcal{G}$. In this case, an ordered Gibbs sampler applied to h, i.e. to the conditional distributions $P(X_1 \mid \mathcal{MB}_1(X_1)), \ldots, P(X_n \mid \mathcal{MB}_n(X_n))$, has a unique stationary distribution $P(\mathbf{X})$, which is exactly the stationary distribution of an ordered Gibbs sampler applied to $h_{\mathcal{G}}$ [236, 113]. The existence of a unique stationary distribution for the ordered Gibbs sampler is warranted by the fact that each conditional distribution $P(X_i \mid \mathcal{MB}_i(X_i))$ is strictly positive. Moreover, since h and $h_{\mathcal{G}}$ define the same probability distribution, h entails that $P(X_i \mid \mathbf{X} \setminus \{X_i\}) = P(X_i \mid \mathcal{MB}_i(X_i))$. This follows from the fact that $P(X_i \mid \mathbf{X} \setminus \{X_i\}) = P(X_i \mid \mathcal{MB}_{\mathcal{G}}(X_i))$, as entailed by Theorem 2.4 (see Section 2.2.2);
2. $P(X_i \mid \mathcal{MB}_i(X_i)) = P(X_i \mid \mathcal{MB}_{\mathcal{G}^*}(X_i))$, where $\mathcal{G}^*$ is the undirected union of $\mathcal{G}_1, \ldots, \mathcal{G}_n$ and $P(X_i \mid \mathcal{MB}_{\mathcal{G}^*}(X_i))$ is derived from a Markov random field $h_{\mathcal{G}^*}$ with graph $\mathcal{G}^*$. In this case, an ordered Gibbs sampler applied to h has a unique stationary distribution $P(\mathbf{X})$, which is exactly the stationary distribution of an ordered Gibbs sampler applied to $h_{\mathcal{G}^*}$ [106, 113]. The existence of a unique stationary distribution for the Gibbs sampler is warranted by the positivity of each conditional distribution $P(X_i \mid \mathcal{MB}_i(X_i))$. Moreover, since h and $h_{\mathcal{G}^*}$ define the same probability distribution, h entails that $P(X_i \mid \mathbf{X} \setminus \{X_i\}) = P(X_i \mid \mathcal{MB}_i(X_i))$. This follows from the fact that $P(X_i \mid \mathbf{X} \setminus \{X_i\}) = P(X_i \mid \mathcal{MB}_{\mathcal{G}^*}(X_i))$, as entailed by Theorem 3.1 (see Section 3.2.1).

We see that, in both cases, the two conditions specified in the theorem (i.e. existence of a unique stationary distribution specified by h and identifiability of a set of conditional independencies entailed by that distribution) are satisfied by the considered hybrid random field. $\square$

We refer to the second condition in Theorem 4.1 as the *modularity property* (or *modularity condition*). Accordingly, we refer to Theorem 4.1 as the *modularity theorem*. Based on the modularity property, the set $\mathcal{MB}_i(X_i)$ is a Markov blanket of X_i in $\mathbf{X}$. If we look at the way that Bayesian networks and Markov random fields are designed, we see that the mathematical foundation of the respective likelihood functions lies in some kind of conditional independence property. While the estimation of joint probabilities in Bayesian networks is allowed by the directed Markov assumption, using Markov random fields requires to assume that the statistical conditions detailed in the Hammersley-Clifford theorem are satisfied. Although independence assumptions may fail to hold in empirical domains, due to the simplifying nature of such assumptions, the resulting models can be quite effective in practice. The conditional independence property one assumes when using hybrid random fields is the modularity condition.

In order to extract a joint probability distribution from a hybrid random field, Gibbs sampling techniques need to be used [113]. A simple and general Gibbs sampling algorithm is described in Section 4.4. Unfortunately, Gibbs sampling can be computationally expensive, which encourages us to look for an alternative way of measuring joint probabilities. Following the strategy employed in Section 3.2.2 for Markov random fields, our idea is to use the pseudo-likelihood function:

$$P^*(\mathbf{X}) = \prod_{i=1}^{n} P(X_i \mid \mathbf{X} \setminus \{X_i\}) \qquad (4.2)$$

Based on the modularity property, Equation 4.2 can be simplified as follows:

$$P^*(\mathbf{X}) = \prod_{i=1}^{n} P(X_i \mid \mathcal{MB}_i(X_i)) \qquad (4.3)$$

Therefore, in order to measure a pseudo-likelihood in a hybrid random field, we only need to be able to compute the conditional distribution of each node X_i given the state $mb_i(X_i)$ of $\mathcal{MB}_i(X_i)$. In Section 2.2.2 we saw how this can be done in a simple and very efficient way.

An important comment to make concerns the possible presence of loops in hybrid random fields. Clearly, loops cannot arise at the local level, that is at the level of each Bayesian network, since this is prevented (trivially) by the very use of Bayesian networks as representations of the local conditional distributions. On the other hand, loops can arise at the global level, since it may happen (as shown in Figure 4.1) that two nodes (such as X_1 and X_3) point to one another if considered simultaneously in different Bayesian networks. This is not a problem at all, because of the modular nature of hybrid random fields. Global loops do not affect the way that the pseudo-likelihood is computed, since that function is factorized as a product of local distributions such that each one of them is computed independently of the remaining ones. Moreover, since the presence of a cycle in the directed union of graphs $\mathcal{G}_1, \ldots, \mathcal{G}_n$ prevents the hybrid random field from satisfying condition 2.a in Definition 4.4, this ensures that condition 2.b is instead satisfied, and hence that the hybrid random field correctly specifies a joint distribution via a suitable Markov random field (with graph $\mathcal{G}^* = \mathcal{G}_1 \uplus \ldots \uplus \mathcal{G}_n$).

4.3 Formal Properties

We now show that hybrid random fields are a more general formalism than Bayesian networks and Markov random fields. In particular, the result we are going to establish is that the class of (conditional) independence structures representable by means of hybrid random fields is given exactly by the union of the classes containing the (conditional) independence structures that can be specified by Bayesian networks and Markov random fields respectively. By *conditional independence structure* (or simply *independence structure*) we mean a set of conditional independence statements concerning a specified collection of random variables. Hence, the conditional independence structure represented by a given graphical model is nothing but the set of conditional independence statements entailed by (the structure of) that graphical model. Since it is known that some independence structures represented by Bayesian networks cannot be represented by Markov random fields and vice-versa [237], this means that both the class of independence structures captured by Bayesian networks and the class of independence structures captured by Markov random fields are strictly included in the class captured by hybrid random fields.

Before establishing the relevant properties, we need to recall one lemma concerning the relationships between Bayesian networks and Markov random fields:

Lemma 4.1. *Let us denote by $\mathcal{BN}$ and $\mathcal{MRF}$ the set of independence structures that are representable by Bayesian networks and Markov random fields respectively. Then, the relationship between $\mathcal{BN}$ and $\mathcal{MRF}$ is characterized by the following properties:*

$$\mathcal{BN} \cap \mathcal{MRF} \neq \emptyset$$
$$\mathcal{BN} \setminus \mathcal{MRF} \neq \emptyset \qquad (4.4)$$
$$\mathcal{MRF} \setminus \mathcal{BN} \neq \emptyset$$

Proof. See e.g. [237]. □

The conditional independence structures that are representable by both Bayesian networks and Markov random fields (i.e. the elements of $\mathcal{BN} \cap \mathcal{MRF}$) are usually referred to as *decomposable models* [103, 57, 20].

We can now prove the first important lemma concerning the generality properties of hybrid random fields:

Lemma 4.2. *If $\mathcal{HRF}$ is the set of independence structures that can be represented by hybrid random fields, then*

$$\mathcal{BN} \subseteq \mathcal{HRF} \qquad (4.5)$$

where $\mathcal{BN}$ is the set of independence structures that are representable by Bayesian networks.

Proof. In order to prove the lemma, we show that, for any Bayesian network $h_{\mathcal{G}}$ with DAG $\mathcal{G} = (\mathcal{V}, \mathcal{E})$, there exists a hybrid random field h representing the same distribution $P(\mathbf{X})$ represented by $h_{\mathcal{G}}$. In other words, we show that any independence structure represented by a Bayesian network can be mapped onto an equivalent independence structure represented by a hybrid random field. Suppose that, for each X_i in $\mathbf{X}$, $\mathcal{MB}_{\mathcal{G}}(X_i)$ is the Markov blanket of X_i in $\mathcal{G}$. Then, for each BN_i in h, set the DAG $\mathcal{G}_i$ to be given by $(\mathcal{V}_i, \mathcal{E}_i)$, where $\mathcal{V}_i = \{X_i\} \cup \mathcal{MB}_{\mathcal{G}}(X_i)$ and $\mathcal{E}_i = \{(X_j, X_k) : \{X_j, X_k\} \subseteq \mathcal{V}_i \wedge (X_j, X_k) \in \mathcal{E}\}$. Moreover, take the conditional probability tables for the nodes in $\mathcal{V}_i$ to be the same as the corresponding tables in $h_{\mathcal{G}}$.

In this case, we have that, for each X_i in $\mathbf{X}$, $P(X_i \mid \mathcal{MB}_i(X_i)) = P(X_i \mid \mathcal{MB}_{\mathcal{G}}(X_i))$, which means that the stationary distribution of an ordered Gibbs sampler applied to h will be exactly $P(\mathbf{X})$. □

Second, we prove an analogous lemma regarding the relationship between Markov and hybrid random fields:

Lemma 4.3. *If* $\mathcal{HRF}$ *is the set of independence structures that can be represented by hybrid random fields, then*

$$\mathcal{MRF} \subseteq \mathcal{HRF} \tag{4.6}$$

where $\mathcal{MRF}$ *is the set of independence structures that are representable by Markov random fields.*

Proof. The lemma is proved by showing that, for any Markov random field $h_{\mathcal{G}^*}$ with graph $\mathcal{G}^* = (\mathcal{V}^*, \mathcal{E}^*)$, there exists a hybrid random field h representing the same joint distribution $P(\mathbf{X})$ represented by $h_{\mathcal{G}^*}$. In other words, we show that any independence structure represented by a Markov network can be mapped onto an equivalent independence structure represented by a hybrid random field. Suppose that, for each X_i in $\mathbf{X}$, $\mathcal{MB}_{\mathcal{G}^*}(X_i)$ is the Markov blanket of X_i in $\mathcal{G}^*$. Then, for each BN_i in h, set the DAG $\mathcal{G}_i$ to be given by $(\mathcal{V}_i, \mathcal{E}_i)$, where $\mathcal{V}_i = \{X_i\} \cup \mathcal{MB}_{\mathcal{G}^*}(X_i)$ and $\mathcal{E}_i = \{(X_j, X_i) : X_j \in \mathcal{MB}_{\mathcal{G}^*}(X_i)\}$. Moreover, let the conditional probability table for node X_i in BN_i store the same values specified by $h_{\mathcal{G}^*}$ for the corresponding conditional probabilities. In this case, we have that, for each X_i in $\mathbf{X}$, $P(X_i \mid \mathcal{MB}_i(X_i)) = P(X_i \mid \mathcal{MB}_{\mathcal{G}^*}(X_i))$. This means that the stationary distribution of an ordered Gibbs sampler applied to h will be exactly $P(\mathbf{X})$. □

Given Lemmas 4.1–4.3, we can finally prove the following theorem:

Theorem 4.2. *Let* $\mathcal{BN}$, $\mathcal{MRF}$, *and* $\mathcal{HRF}$ *denote the sets of conditional independence structures that are representable by Bayesian networks, Markov random fields, and hybrid random fields respectively. Then,* $\mathcal{HRF}$ *can be characterized as follows:*

$$\mathcal{HRF} = \mathcal{BN} \cup \mathcal{MRF} \tag{4.7}$$

Proof. Lemmas 4.2–4.3 together imply that

$$\mathcal{BN} \cup \mathcal{MRF} \subseteq \mathcal{HRF} \tag{4.8}$$

On the other hand, it is easily seen that, if $P(\mathbf{X})$ is the probability distribution represented by a hybrid random field h (with DAGs $\mathcal{G}_1, \ldots, \mathcal{G}_n$), then $P(\mathbf{X})$ must be identical to the probability distribution specified by either a Bayesian network $h_{\mathcal{G}}$ (with DAG $\mathcal{G} = \bigcup_{i=1}^n \mathcal{G}_i$), or a Markov random field $h_{\mathcal{G}^*}$ (with graph $\mathcal{G}^*$ given by the undirected union of $\mathcal{G}_1, \ldots, \mathcal{G}_n$), or both (when $P(\mathbf{X})$ is a decomposable model). This derives from the fact that the existence and uniqueness of $P(\mathbf{X})$ results from its being identical to the distribution represented by a Bayesian or Markov network, as made clear by the proof of Theorem 4.1. We can then state the following equality:

$$\mathcal{HRF} \subseteq \mathcal{BN} \cup \mathcal{MRF} \tag{4.9}$$

Therefore, we can conclude that $\mathcal{HRF} = \mathcal{BN} \cup \mathcal{MRF}$, as implied by expressions 4.8–4.9 together. $\square$

An intuitive way of rephrasing the result established by Theorem 4.2 is by saying that the modularity property is mathematically equivalent to the logical disjunction of (*i*) the directed Markov property (holding for Bayesian networks) and (*ii*) the local/global Markov property (holding for Markov random fields).

It is worth noticing that, as a trivial consequence of Lemmas 4.1–4.3, the following inequalities hold:

$$\mathcal{HRF} \setminus \mathcal{BN} \neq \emptyset$$
$$\mathcal{HRF} \setminus \mathcal{MRF} \neq \emptyset \tag{4.10}$$

This means that expressions 4.5–4.6 can be refined as follows:

$$\mathcal{BN} \subset \mathcal{HRF}$$
$$\mathcal{MRF} \subset \mathcal{HRF} \tag{4.11}$$

The fact that hybrid random fields generalize both Bayesian and Markov networks does not mean that hybrid random fields are a 'better' probabilistic model in any absolute sense. In particular, Theorem 4.2 does not provide a general argument for using the hybrid model in preference to traditional directed or undirected graphical models. The proper conclusion we can draw from Theorem 4.2 is that by using hybrid random fields instead of Bayesian networks or Markov random fields we do not lose any representational power. On the contrary, by using the hybrid model we retain the capability of representing any kind of conditional independency that can be represented by either Bayesian or Markov networks. An empirical argument to the effect that hybrid random fields (as estimated by means of a particular structure learning technique) are often able to deliver more accurate predictions than Bayesian networks and Markov random fields in a variety of applications is provided by the results of the experimental demonstrations reported in Chapter 6.

4.4 Inference

Given a set of random variables $X_1, \ldots, X_n$, probabilistic inference is the task of estimating the posterior distribution of a variable X_i given that some other variables have a known value. The variables whose values are known are called *evidence variables*, whereas the *query variable* is the one for which we want to estimate the conditional distribution given the evidence. We now present the ordered Gibbs sampling algorithm [269, 113], which is an approximate inference technique for probabilistic graphical models belonging to the family of Markov chain Monte Carlo (MCMC) methods [113]. We choose to present this version of MCMC-based inference because it is particularly simple to implement, and it can be applied equally well to Bayesian networks, Markov random fields, and hybrid random fields. A more complex inference technique for graphical models, that has been attracting a lot of interest in recent years, is generalized belief propagation [322], which extends to cyclic graphical structures the belief propagation method originally proposed in [237]. Overviews of other inference methods can be found e.g. in [226] and [30].

The basic idea behind MCMC is to run a stochastic simulation over the graphical model for a certain number of cycles. During the simulation, the values of the evidence variables are kept fixed, while the values of the other variables are repeatedly sampled, in turn, according to their conditional distributions given the respective Markov blankets. Before the simulation begins, the values of non-evidence variables are initialized randomly. At each cycle, the value assumed by the query variable is recorded, so that once the simulation has finished, the relative frequencies of the different values assumed by that variable can be used to compute the respective conditional probabilities. A complete description of ordered Gibbs sampling is provided by Algorithm 4.1, where $\mathcal{D}_{X_i}$ is used to denote the domain of variable X_i and $P(X_i \mid mb(X_i); h)$ refers to the conditional distribution of X_i given the state of its Markov blanket, as derived from graphical model h.

Algorithm 4.1 GSInference: Gibbs sampling-based inference for probabilistic graphical models

Input: Probabilistic graphical model h with nodes $X_1, \ldots, X_n$; vector $\mathbf{e}$ containing the values $e_1, \ldots, e_m$ of evidence variables $E_1, \ldots, E_m$, where $\{E_1, \ldots, E_m\} \subset \{X_1, \ldots, X_n\}$; query variable $X_q \in \{X_1, \ldots, X_n\} \setminus \{E_1, \ldots, E_m\}$; number c of cycles to run.
Output: Approximate estimate of each conditional probability $P(x_{q_i} \mid e_1, \ldots, e_m)$, for $1 \leq i \leq |\mathcal{D}_{X_q}|$.

```
GSInference(h, e, X_q, c):
 1.      ε = {E₁,...,Eₘ}
 2.      for(i = 1 to n)
 3.          if(∃j Eⱼ ∈ ε ∧ Xᵢ = Eⱼ)
 4.              Xᵢ = eⱼ
 5.          else
 6.              Xᵢ = random value from 𝒟ₓᵢ
 7.      for(i = 1 to |𝒟ₓ_q|)
 8.          pᵢ = 0
 9.      for(i = 1 to c)
10.          for(j = 1 to n)
11.              if(Xⱼ ∉ ε)
12.                  Xⱼ = random value xⱼₖ sampled from P(Xⱼ | mb(Xⱼ); h)
13.                  if(Xⱼ = X_q)
14.                      pₖ = pₖ + 1
15.      for(i = 1 to |𝒟ₓ_q|)
16.          pᵢ = pᵢ/c
17.      return (p₁,...,p_{|𝒟ₓ_q|})
```

The main theoretical property of MCMC inference is that, as the number of cycles increases, the stochastic simulation carried out by the algorithm settles into a kind of equilibrium such that the (normalized) fraction of cycles that the system spends in a particular state converges to the posterior probability of that state given the evidence, provided that the conditional distribution of each variable given its Markov blanket is strictly positive [269, 113]. This means that, as the simulation proceeds, the estimated distribution of the query variable given the evidence converges to the true posterior distribution of that variable.

4.5 Parameter Learning

In hybrid random fields, parameter learning is the problem of learning the local distributions of the model variables given the respective Markov blankets. That is, for each variable X_i, the task is to learn the parameters of the conditional distribution $P(X_i \mid mb(X_i))$. This requires that the structure of the hybrid random field has been previously fixed, i.e. that the DAG $\mathcal{G}_i$ associated with each variable X_i is known. In order to estimate the model parameters from a dataset $\mathbf{D}$, we simply learn the parameters of each Bayesian network BN_i from $\mathbf{D}$, using the method described in Section 2.3.

4.6 Structure Learning

Structure learning in HRFs is the problem of learning, for each variable X_i, what other variables appear as nodes in BN_i, and what edges are contained in the DAG $\mathcal{G}_i$. Clearly, this means learning the structure of each Markov blanket $\mathcal{MB}_i(X_i)$ within the HRF. While parameter learning assumes that the MB of each variable has already been fixed, the aim of structure learning is to identify each MB and to determine its graphical structure.

We propose a heuristic structure learning algorithm for HRFs, which we call *Markov Blanket Merging* (MBM). MBM works under the hypothesis that all variables are fully observed within the dataset. That is to say, the data do not contain incomplete patterns. The aim of MBM is to find an assignment of MBs $\mathcal{MB}_1(X_1), \ldots, \mathcal{MB}_n(X_n)$ to the nodes $X_1, \ldots, X_n$ that maximizes the model pseudo-likelihood given a dataset $\mathbf{D}$. The basic idea behind MBM is to start from a certain assignment of relatives to the model variables, to learn the local BNs in the model, and then to iteratively refine the assignment so as to come up with MBs that increase the model pseudo-likelihood with respect to the previous assignment. This iterative procedure stops when no further refinement of the MBs assignment increases the value of the pseudo-likelihood. In other words, MBM is a local search algorithm exploring a space of possible MB assignments to the model variables. As such, MBM only warrants convergence to local optima of the pseudo-likelihood function. However, the results achieved in our applications show that the optima reached by

MBM are usually good enough to produce comparatively accurate predictions. The reason why we choose to maximize the pseudo-likelihood, rather than the likelihood in the strict sense, is that the former function is much more efficient to compute than the latter, which requires instead Gibbs sampling.

An important problem to consider is given by the size of the search space. If we allowed the search space to contain all possible MB assignments, the size of the space would be intractable: if n is the number of variables, for each variable there are 2^{n-1} possible MBs, and therefore the size of the search space is $n \cdot 2^{n-1}$. Clearly, exploring such a state space exhaustively is not feasible. For this reason, MBM reduces the size of the search space by assuming that the cardinality of each MB within the HRF is bounded by an upper limit k^* (where the particular value of the k^* parameter is tuned in each application by preliminary cross-validation). While the usefulness of this assumption will be made clear by the formal analysis of Section 4.6.4, the accuracy of the models learned in various applications (Chapter 6) shows that the assumption is quite reasonable in practice.

In order to develop the algorithm, we specify three components: (*i*) a model initialization strategy, that is a way to produce the initial assignment of Markov blankets to the model variables; (*ii*) a search operator, that is a way to refine a given assignment so as to produce an alternative assignment; (*iii*) an evaluation function, that is a way to evaluate a given assignment. These three components are described in Sections 4.6.1–4.6.3.

4.6.1 Model Initialization

The way MBM produces an initial assignment is by choosing an initial size k of the sets of relatives, and then by assigning as relatives to each variable X_i those k variables that achieve the highest scores on the χ^2 dependence test with respect to X_i. The intuitive motivation for this choice is that a set of relatives $\mathcal{R}(X_i)$ containing variables that are more strongly correlated to X_i is more likely to capture the MB of X_i than a set of relatives containing variables that are only weakly correlated to X_i. Given the sets of relatives $\mathcal{R}(X_1), \ldots, \mathcal{R}(X_n)$, an assignment of MBs to the variables $X_1, \ldots, X_n$ is obtained by learning (both the structure and the parameters of) a BN BN_i for each X_i, where BN_i contains X_i together with $\mathcal{R}(X_i)$. In order to learn the structure of the local Bayesian networks, we use Algorithm 2.1. In each application, the value of k can be tuned by preliminary cross-validation.

4.6.2 Search Operator

Given a current assignment of MBs to the model variables, where the assignment is given by the MBs $\mathcal{MB}_1(X_1), \ldots, \mathcal{MB}_n(X_n)$ specified by the networks $BN_1, \ldots, BN_n$, a new assignment is obtained as follows. For each variable X_i (where $1 \le i \le n$), we construct the set $\mathcal{U}_i$ as the union of $\mathcal{MB}_i$ with the Markov blankets of X_i in all graphs $\mathcal{G}_j$ such that X_i appears in $\mathcal{G}_j$ within the current assignment. Given the sets $\mathcal{U}_1,\ldots,\mathcal{U}_n$, we first check whether the cardinality of each $\mathcal{U}_i$ does not exceed a certain threshold k^*, and then we construct a new set of Bayesian networks $BN_1^*,\ldots,BN_n^*$ such that, for each i, if $|\mathcal{U}_i| \leq k^*$, then BN_i^* is the network learned by using the set $\mathcal{R}(X_i) = \mathcal{U}_i$ as the new set of relatives of X_i, whereas, if $|\mathcal{U}_i| > k^*$, then $BN_i^* = BN_i$. Given $BN_1^*,\ldots,BN_n^*$, for each X_i we compare the value $\sum_{j=1}^{|\mathbf{D}|} \log P\left(x_{i_j}|mb_{i_j}(X_i)\right)$ to the value $\sum_{j=1}^{|\mathbf{D}|} \log P\left(x_{i_j}|mb_{i_j}^*(X_i)\right)$. These values are, respectively, the conditional log-likelihoods of X_i given its Markov blanket, as determined on the one hand by BN_i and on the other hand by BN_i^*. If the latter value is higher than the former, that is if the conditional log-likelihood of X_i given its MB increases after replacing $\mathcal{MB}_i(X_i)$ with $\mathcal{MB}_i^*(X_i)$, then $\mathcal{MB}_i^*(X_i)$ is chosen as the Markov blanket of X_i in the new assignment, otherwise X_i is assigned again $\mathcal{MB}_i(X_i)$. As for the parameter k, a suitable value for $k*$ can also be determined by means of cross-validation.

An important point to note is that, at each iteration of MBM, the MBs of the variables are replaced by the new ones simultaneously at the end of the cycle, and not incrementally. This choice makes MBM insensitive to the ordering of the variables, which is also apparent from Algorithm 4.2.

4.6.3 Evaluation Function

An assignment of MBs to the variables is evaluated by measuring the model pseudo-log-likelihood. That is, our evaluation function for a model h given dataset $\mathbf{D}$ will be the following:

$$\log P^*(\mathbf{D} \mid h) = \sum_{j=1}^{m} \sum_{i=1}^{n} \log P\left(X_i = x_{i_j} \mid mb_{i_j}(X_i)\right) \tag{4.12}$$

where $m = |\mathbf{D}|$. Actually, by building an alternative assignment from the current one we implicitly evaluate the new assignment. In fact, the new assignment will differ from the old one only if there is at least one variable X_i such that the new MB of X_i increases the conditional log-likelihood of X_i given the MB. Based on the definition of the pseudo-log-likelihood function and the way the search operator works, an increase in any one of the n local log-likelihoods of the model ensures an increase in the global pseudo-log-likelihood. The reason is that the search operator works in a modular fashion: the way each $\mathcal{MB}_i(X_i)$ is modified by the operator is such that the change does not affect any other MB in the model. Therefore, after we build a new assignment, it is sufficient to compare it to the old one in order to know whether it increases the model pseudo-likelihood: if the two assignments are different, then we can endorse the new one as being better, otherwise we keep the old one and stop the search. A detailed description of MBM is provided by Algorithm 4.2, where $\chi^2(X_i,X_j,\mathbf{D})$ denotes the value of the χ^2 statistic, as computed from dataset $\mathbf{D}$ for variables X_i and X_j.

Algorithm 4.2 MBMLearnHRF: MBM structure learning in hybrid random fields

Input: Dataset $\mathbf{D}$, containing m observations for the random variables $X_1, \ldots, X_n$; integers k, k^*.
Output: Hybrid random field h (locally) maximizing the heuristic function $\log P^*(D \mid h)$.

MBMLearnHRF($\mathbf{D}, k, k^*$):
```
 1.    𝒳 = {X₁,…,Xₙ}
 2.    for(i = 1 to n)
 3.        ℛᵢ = {Xⱼ : Xⱼ ∈ 𝒳 \ {Xᵢ} ∧ |{Xₖ : χ²(Xᵢ,Xₖ,D) > χ²(Xᵢ,Xⱼ,D)}| < k}
 4.    do
 5.        assignmentWasRefined = false
 6.        for(i = 1 to n)
 7.            𝒱 = {Xᵢ} ∪ ℛᵢ
 8.            BNᵢ = a Bayesian network with DAG 𝒢 = (𝒱, ∅)
 9.            BNᵢ = HCLearnBN(D, BNᵢ)    //See Algorithm 2.1
10.        for(i = 1 to n)
11.            𝒰ᵢ = ⋃ⱼ₌₁ⁿ ℳℬⱼ(Xᵢ)
12.            if(|𝒰ᵢ| ≤ k*)
13.                𝒱 = {Xᵢ} ∪ 𝒰ᵢ
14.                BNᵢ* = a Bayesian network with DAG 𝒢 = (𝒱, ∅)
15.                BNᵢ* = HCLearnBN(D, BNᵢ*)    //See Algorithm 2.1
16.                ℳℬᵢ = the MB of Xᵢ in BNᵢ
17.                ℳℬᵢ* = the MB of Xᵢ in BNᵢ*
18.                if(∑ⱼ₌₁ᵐ log P(xᵢⱼ | mbᵢⱼ*(Xᵢ)) > ∑ⱼ₌₁ᵐ log P(xᵢⱼ | mbᵢⱼ(Xᵢ)))
19.                    ℛᵢ = ℳℬᵢ*(Xᵢ)
20.                    assignmentWasRefined = true
21.    while(assignmentWasRefined)
22.    return {BN₁,…,BNₙ}
```

Figure 4.2 illustrates the way MBM iteratively refines the structure of a HRF, displaying an assignment of MBs at time t, and the assignment obtained at time $t + 1$. MBM compares the model pseudo-likelihood at time $t + 1$ to the pseudo-likelihood at time t: if the pseudo-likelihood is higher at $t + 1$, the new set of BNs is retained and another iteration of MBM is run, otherwise the algorithm stops searching and the set of BNs obtained at time t is returned.

4.6.4 Discussion

Given the design of the algorithm, MBM does not warrant the model pseudo-likelihood to converge to the global optimum of the function. The main reason for this is given by the choice of the two parameters k and k^*, fixing respectively an initial value and an upper bound for the size of the sets of relatives to be considered for each node. However, MBM does warrant that the model pseudo-likelihood will only increase during the learning process, since any candidate model that decreases the scoring function will be discarded during the search. This feature of MBM marks

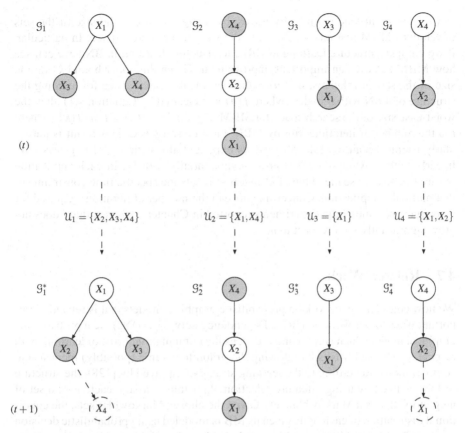

Fig. 4.2 An iteration of MBM. Given the assignment at time t, a new assignment is obtained by merging the (possibly different) Markov blankets of each variable within the different Bayesian networks, and then by learning a new Bayesian network for each $\{X_i\} \cup \mathcal{U}_i$

an advantage over other efficient algorithms for Markov blanket discovery, such as the one proposed in [148].

One very important point to consider is the following one. As it should be apparent from the formulation of the MBM algorithm given above, MBM does not warrant that the returned model is a hybrid random field in the strict sense. In particular, the output of the algorithm may not satisfy condition 2 in the definition of HRF (see Definition 4.4 in Section 4.2). However, as the experimental analysis presented in Chapter 6 will show, this theoretical limitation of MBM does not lead to serious troubles, as far as we are concerned with the practical behavior of the learned HRFs, since the models estimated through Markov Blanket Merging display a comparatively high prediction accuracy in all considered applications. On the other hand, we also believe that it might be an important research challenge the idea of developing structure learning algorithms which are able to warrant the estimated model to be a HRF in the strict sense.

Concerning instead complexity issues, by using bounds on the size of the sets of relatives MBM restricts significantly the size of the search space. In particular, if we compare structure learning in HRFs to structure learning in BNs, we can see how MBM achieves an important improvement. Given the upper bound k^* on the size of the sets of relatives, if $f(n)$ is the size of the search space for learning the structure of a BN with n nodes (where $f(n)$ is measured by Equation 4.1), then the worst-case size of the search space for MBM is given by $f^*(n) = i \cdot n \cdot f(k^*)$, where i is the number of iterations run by MBM until convergence. This result is particularly useful, because while $f(n)$ grows exponentially with n, $f^*(n)$ grows only linearly with n. Although $f^*(n)$ grows exponentially with k^*, in each application we are free to choose the value of k^* that best meets the specific time constraints of that particular application. Concerning instead the number of iterations required for convergence, some of the experiments reported in Chapter 6 suggest that i does not grow significantly with respect to n.

4.7 Related Work

We now consider some work on probabilistic graphical models that is related in important ways to our work on HRFs. Dependency networks (DNs) are a probabilistic graphical model which significantly reduces the computational cost of learning with respect to BNs and MRFs by allowing the graph to assume (possibly) inconsistent configurations. According to the learning strategy employed by [128], the structure of DNs is fixed by using a feature selection algorithm to assign each node a set of neighbors (that is a Markov blanket). Given the chosen Markov blankets, the conditional distribution of each node given its MB is modeled using probabilistic decision trees [100, 38]. One difference between hybrid random fields and dependency networks is that using BNs as models of the local conditional distributions allows us to develop an iterative structure learning algorithm. In Markov Blanket Merging, feature selection only initializes the model structure, and the model can then be refined to better fit the data distribution. Given the way that dependency networks are defined in [128], hybrid random fields may be viewed as a particular class of 'general' dependency networks,[2] designed with the primary aim of allowing for iterative structure learning. In this respect, as the experimental demonstration presented in Chapter 6 will show, the effects of introducing this kind of structure learning within the framework defined by DNs are quite encouraging, since it leads to dramatic improvements in terms of the prediction accuracy of the learned models. On the other hand, consistent dependency networks are a less general model than hybrid random fields, since the equivalence of (consistent) dependency networks and Markov random fields implies that the class of conditional independence structures

---

[2] According to the terminology used in [128], general dependency networks are DNs that are not necessarily 'consistent', while consistent dependency networks (i.e. dependency networks in the strict sense) are defined in such a way that they turn out to be equivalent to Markov random field models.

representable by DNs is strictly included in the class of independence structures representable by HRFs (see Section 4.3).

Graphical chain models [191, 314], or chain graphs, define a mathematical framework for studying the properties of probabilistic graphical models in a mostly general way. Bayesian networks and Markov random fields are known to be special cases of chain graphs. While the basic idea of developing a formalism capable of subsuming some classes of known probabilistic graphical models is common to chain graphs and hybrid random fields, our work on the latter models is more strictly focused on learning and its algorithmic aspects, such as scalability. In practical applications, chain graphs have been estimated typically in the form of particular graphical models, such as Bayesian and Markov networks, or combinations thereof [42]. In this respect, since in experimental evaluations hybrid random fields have been compared to various kinds of Bayesian networks and Markov random fields [95, 94, 97], they have been compared thereby to the most representative kinds of chain graphs for which effective learning algorithms exist [39, 40].

A computationally efficient method for learning MBs is the Max-Min Markov Blanket (MMMB) algorithm [305]. A crucial difference between the MMMB method and MBM is that the former method is committed to the faithfulness assumption with respect to the estimated Bayes nets, whereas Markov Blanket Merging dispenses with any such assumption. While assuming faithfulness may be reasonable in some applications, that condition cannot be assumed to hold in general [226]. Therefore, in this respect MBM is a more general Markov blanket learning technique than MMMB.

The *local-to-global-search* algorithm [148] for Markov blanket discovery achieves computational efficiency without making the faithfulness assumption. However, this algorithm is not generally warranted to optimize the chosen evaluation function, since the score of the learned network can happen to decrease during the learning process.

4.8 Final Remarks

Probabilistic graphical models have long been classified, broadly, into directed and undirected models. This chapter presented to the reader a new kind of graphical model, namely hybrid random fields, which inherit properties of both directed and undirected graphical models. Although the formal representation of hybrid random fields exploits (a collection of) directed acyclic graphs, the hybrid nature of these models results from their capability to model with equal flexibility both the conditional independence structures representable by means of Bayesian networks and the independence structures representable by Markov random fields. In particular, an appealing mathematical property of HRFs is that the class of independence structures they are able to represent is given by the union of the sets of structures representable by Bayesian and Markov networks, respectively. As we saw, the key to identifying the set of independence statements entailed by any given HRF lies in the so-called modularity property, which was proved to hold for any hybrid random field by means of Theorem 4.1. This theorem is of crucial importance for the mathematical framework underlying HRFs, because on the one hand it warrants that the model provides a well-defined representation of joint probability distributions, and on the other hand it allows to map any represented distribution onto an explicit set of conditional independence statements characterizing the relationships between the modeled random variables.

An interesting point to consider is the following. In their original presentation [95, 94, 97], hybrid random fields were defined as a probabilistic graphical model which *assumes* the modularity condition as a defining feature. This allowed to establish a couple of (somewhat loose) relationships of the model with Bayesian and Markov networks, but it did not allow to state an exhaustive characterization of the conditional independence properties holding for the joint probability distributions represented by hybrid random fields. On the contrary, this chapter provided a more restrictive definition of the model, which allows on the one hand to prove (through Theorem 4.1) that the modularity condition holds for any HRF, and on the other hand to provide a tighter characterization of the relationships linking the class of independence structures representable by HRFs to the classes of structures representable by BNs and MRFs respectively. In a concise way, we may characterize these relationships by stating that the probabilistic property referred to as modularity condition is mathematically equivalent to the logical disjunction of the two probabilistic properties referred to as (*i*) directed Markov assumption and (*ii*) local/global Markov property.

After presenting the hybrid random field model and investigating its theoretical properties, we presented algorithms for performing (approximate) inference over HRFs (Section 4.4), and for learning not only their parameters (Section 4.5), but also their structure from data (Section 4.6). With respect to structure learning, we need to stress the fact that the provided algorithm (i.e. the Markov Blanket Merging technique) is not warranted to return a HRF in the strict sense, i.e. the output of the algorithm may not satisfy the second condition in the definition of HRF (see Section 4.2). As the experimental investigation presented in Chapter 6 will show, this limitation of MBM is not so serious as far as we are interested in the practical effects of using it, since the models estimated through Markov Blanket Merging display a comparatively high prediction accuracy in all considered applications. On the other hand, we think it might be an important motivation for future research the goal of developing structure learning algorithms which may warrant the estimated model to strictly satisfy the definition of hybrid random field. Finally, we briefly discussed some other probabilistic graphical models and learning techniques related to hybrid random fields (Section 4.7), so as to help the reader to correctly locate the hybrid model within the wide and rich scenario displayed by the literature on graphical models.

Chapter 5
Extending Hybrid Random Fields: Continuous-Valued Variables

"Aus dem Paradies, das Cantor uns geschaffen, soll uns niemand vertreiben können."

David Hilbert, 1926 [134]

5.1 Introduction

In Section 1.2 we introduced the concept of feature, or attribute. The idea is that, in order to apply statistics or learning machines to phenomena occurring in the real world, it is necessary to describe them in a proper feature space, such that each phenomenon can be thought of as a random variable (if a single attribute is used), or a random vector (if multiple features are extracted). The feature extraction process is crucial for the success of an application, since there is no good model that can compensate for wrong, or poor, features. Although several types of attributes are sometimes referred to in the literature (e.g., categorical, nominal, ordinals, etc.), two main families of features can be pointed out, namely discrete and continuous-valued (or, simply, continuous). Attributes are discrete when they belong to a domain which is a countable set, such that each feature can be seen as a symbol from a given alphabet. Examples of discrete features spaces are any subsets of the integer numbers, or arbitrary collections of alphanumeric characters. Continuous-valued feature spaces are compact subsets of $\mathbb{R}$—in which case the attribute is thought of as the outcome of a real-valued random variable—or $\mathbb{R}^d$ for a certain integer $d > 1$—in which case the d-dimensional feature vector is assumed to be the outcome of a real-valued random vector.

The type of the attributes (either discrete or continuous) may depend on the choices made by the designer of the system, i.e. on the specific feature extraction process, and/or on the very nature of the phenomena under consideration. For instance, probabilistic models of sequences of amino-acids within the primary structure of eukaryote proteins are suitably represented via discrete attributes, since each amino-acid in the protein sequence belongs to a closed set of twenty possible standard alternatives (that biologists readily represent either with 1-letter or 3-letter unique codes). Also, boolean (i.e., binary) data are inherently discrete. On the other hand, most physical measurements are likely to be expressed as continuous values, e.g. the weight of a person or of a molecule, the distance between two subatomic

A. Freno and E. Trentin: Hybrid Random Fields, ISRL 15, pp. 87–119.
springerlink.com   © Springer-Verlag Berlin Heidelberg 2011

particles, the energy of a system. Nonetheless, there may be ambiguous situations. The distance between two cities is, speaking in physical terms, a continuous value (unless a discrete-space model of the Universe is adopted, which is supported by several theories in modern physics). But it is reasonable to assume that a rounding is carried out, turning up disentangling, yet equally useful to all practical effects (nobody cares that the distance is 147.24 miles rather than 147.31, it is 'just' 147 miles). In other words, an implicit *discretization* of the otherwise naturally continuous values is applied. Notwithstanding the discretization, eventually a real-valued variable can, once again, turn out to be a better representation of city distances: if a discrete attribute entails an indeed unbearable number of distinct outcomes for any possible distance in miles $(\ldots, 1043, 1044, 1045, \ldots)$, then it is clearly not viable in any practical model. Contrariwise, phenomena that are intrinsically continuous-valued, like a visual scene in front of a camera, may be better represented sometimes in terms of a dense grid of binary pixels, i.e. as a collection of discrete attributes.

5.1.1 An Abiding Debate: Discrete vs Continuous Variables

The vast majority of the literature on graphical models assumes discrete random variables—if not in the theoretical analysis, typically in the design and implementation of the model. Several concepts involved even in the definition of the different paradigms require a discrete representation of the information. For instance, the popular definitions of Bayesian networks and Markov random fields we gave in the previous chapters rely on such notions as conditional probability table or feature function, respectively (although the corresponding definitions may be modified so as to cover continuous features, as we will see throughout this chapter). Extension to continuous-valued attributes is often left implicit, and no unique standards have emerged so far (different authors proposed alternative approaches). In many cases, it is simply assumed that discrete features are sufficient, i.e. that (*i*) a discretization of real-valued variables can be systematically accomplished, and (*ii*) such a discretization does not compromise the performance of the resulting machine. Although the representation of information in the memory of a digital computer is intrinsically discrete (at a certain micro-level of granularity), the topic of whether or not continuous-valued models offer concrete advantages over their discrete counterparts is still a debated issue. For example, the work presented in [321] advocates the edge of discretized on continuous variables, showing that the former may even lead to improved performance w.r.t. the latter in a naive Bayes classifier, and giving a mathematical proof of conditions under which such a discrete classifier yields the same probability distribution that is obtained from the continuous values directly. Furthermore, discrete models are usually faster and much simpler, being candidates for increased generalization capabilities. As a matter of fact, empirical evidence is reported in [202] which shows that the discretization of continuous data may result in improved classification performance in some genomics and proteomics tasks, when the classifier (e.g., support vector machine, naive Bayes) is affected by the high dimensionality of the feature space.

On the other hand, neglecting the relevance of information brought in by continuous features would implicitly downgrade most of the classic approaches to machine learning and statistical pattern recognition (which have been building sophisticated models on real-valued feature vectors for more than half a century). In so doing, multinomial probability distributions (relying on histograms, relative frequencies, and conditional probability tables) would be enough, indeed, for representing any random quantities. On the contrary, continuous models assume that relevant information is conveyed by real-valued data, information which is gone once discretization, in any given form, is applied. Actually, a certain information loss is a factual consequence of the discretization process, not a mere assumption. Moreover, since a simple rounding of real numbers is often way too brutal and disrespectful of the nature of the data, discretization requires the application of non-trivial algorithms. No standards exist in this respect. The community has been developing more and more effective techniques throughout the years, implicitly pinpointing the fact that discretization of random variables is not as obvious as it might seem at a first glance. Sophisticated techniques have been developed for partitioning real-valued domains into discrete *regions* (e.g., intervals) that can best capture the information which is relevant to density estimation, classification, or regression tasks. For a sample of such approaches and their experimental comparisons see e.g. [43, 85, 267]. Clustering techniques [73] are applied, as well. Their aim is to partition the feature space into separate clusters having a high internal concentration of data (as measured on a sample dataset). The 'centroid' of each cluster can then be elected as the prototypical representative in charge of all possible continuous values belonging to the corresponding cluster (eventually, the feature space is made discrete in terms of a finite collection of such prototypes). We will say more on clustering—in passing—in Section 5.3.2, where the most popular clustering procedure (the k-means algorithm) emerges as the consequence of the maximum-likelihood technique applied to the estimation of the parameters of a Gaussian mixture model under specific assumptions. Partitioning the feature space into either regions, intervals, or clusters, has the unpalatable side-effect of inducing a distortion of the topological nature of the original space. At the end of the day, whatever discretization algorithm you choose, it is going to return a non-uniquely defined solution, entailing a certain information loss, and which takes time to be run (i.e., there is a trade-off in terms of computational burden between the simplicity of the probabilistic model and the complexity of the discretizer).

Empirical evidence of the relevance of continuous variables over discrete attributes has been brought to the attention of the community, as well. For instance, the study reported in [208] on a vowel discrimination task involving the representation and modeling of acoustic signals shows that the use of continuous-valued features allows for the development of a much better predictor than discrete attributes do. Automatic speech recognition systems relying on continuous-density hidden Markov models are known to perform usually better than recognizers based on discrete HMMs, too [301].

5.1.2 Optimal Convergence to the Real Probability Density Function: The Parzen Window Estimator

Let us work out a significant example, whose meaning to the scope of this chapter is twofold. On the one hand, it should make definitely clear the relevance of continuous-value modeling. On the other hand, it offers a theoretical and algorithmic background to the non-parametric density estimation techniques that are developed later in the framework of graphical models. The example concerns the popular Parzen window (PW) method [234]. The fundamentals of the following discussion are based mostly on the excellent, evergreen treatment of the subject handed out by Richard Duda and Peter Hart in 1973 [72]. Let us consider a probability density function (pdf) $p(\mathbf{x})$, defined over a real-valued, d-dimensional feature space. The probability that (the generic outcome of) a random vector $\mathbf{x}^* \in \mathbb{R}^d$, drawn from $p(\mathbf{x})$, falls in a given region $\mathcal{R}$ of the feature space is $P = \int_{\mathcal{R}} p(\mathbf{x}) \, d\mathbf{x}$. Let then $\mathcal{T} = \{\mathbf{x}_1, \ldots, \mathbf{x}_n\}$ be a sample of n outcomes, which are independent and identically distributed (iid) according to $p(\mathbf{x})$. If k_n outcomes in $\mathcal{T}$ fall within $\mathcal{R}$, an empirical frequentist estimate of P can be obtained as $P \simeq k_n/n$. If $p(\mathbf{x})$ is a continuous function, and $\mathcal{R}$ is small enough to prevent $p(\mathbf{x})$ from varying its value over $\mathcal{R}$ in a significant manner, we are also allowed to write $\int_{\mathcal{R}} p(\mathbf{x}) \, d\mathbf{x} \simeq p(\mathbf{x}^*) V$, where $\mathbf{x}^* \in \mathcal{R}$, and V is the volume of region $\mathcal{R}$. As a consequence of the discussion, we can obtain an estimated value of the pdf $p(\mathbf{x})$ over the generic vector $\mathbf{x}^*$ as:

$$p(\mathbf{x}^*) \simeq \frac{k_n/n}{V_n} \tag{5.1}$$

where V_n denotes the volume of region $\mathcal{R}_n$ (i.e., the choice of the region width is explicitly written as a function of n), assuming that smaller regions around $\mathbf{x}^*$ are considered as the sample size n increases. This is expected to allow Equation 5.1 to yield improved estimates of $p(\mathbf{x})$, i.e. to converge to the exact value of $p(\mathbf{x}^*)$ as n (hence, also k_n) tends to infinity. (A thorough discussion of the asymptotic behavior of nonparametric models of this kind can be found in [72], and it will be summarized shortly).

The basic instance of the Parzen window technique assumes that $\mathcal{R}_n$ is a hypercube having edge h_n, such that $V_n = h_n^d$. The edge (or 'bandwidth') h_n can be properly defined as a function of n as $h_n = h_1/\sqrt{n}$, in order to ensure a correct asymptotic behavior. The value h_1 has to be chosen empirically, and it affects the resulting model. The formalization of the idea requires first to define a unit-hypercube window function φ, centered in the origin, in the form

$$\varphi(\mathbf{y}) = \begin{cases} 1 & \text{if } |y_j| \leq \frac{1}{2}, \ j = 1, \ldots, d \\ 0 & \text{otherwise} \end{cases} \tag{5.2}$$

such that $\varphi(\frac{\mathbf{x}^* - \mathbf{x}}{h_n})$ has value 1 iff $\mathbf{x}^*$ falls within the d-dimensional hyper-cubic region R_n centered in $\mathbf{x}$ and having edge h_n. This implies that $k_n = \sum_{i=1}^n \varphi(\frac{\mathbf{x}^* - \mathbf{x}_i}{h_n})$. Using this expression, from Equation 5.1 we can write

$$p(\mathbf{x}^*) \simeq \frac{1}{n} \sum_{i=1}^{n} \frac{1}{V_n} \varphi(\frac{\mathbf{x}^* - \mathbf{x}_i}{h_n}) \tag{5.3}$$

which is the PW estimate of $p(\mathbf{x}^*)$ from the sample $\mathcal{T}$. The model is then refined by considering *smooth* window functions, instead of hypercubes, such as the standard Gaussian kernel with zero mean and unit covariance matrix. This can be shown (theoretically, as well as empirically) to yield much smoother and more realistic estimates of the actual pdf. It has to be emphasized that the Parzen window estimates joint pdfs. The PW approach can also be extended to conditional pdf estimation. In this case, the PW takes the form of another popular non-parametric model, known as the Nadaraya-Watson estimator, which is presented and discussed in Section 5.5.1.

As anticipated, all the quantities involved in the calculation of the PW estimate are indexed with the subscript n, meaning that they are a function of the size of the available dataset. This is crucial, for at least two reasons. First, it points out that, as one would reasonably expect, the regions shall adapt according to the data. In particular, the larger the amount of data and their concentration, the smaller the regions (and, the higher the precision of the estimate). Second, as discussed in [72], in so doing the asymptotic behavior of the model can be investigated as n increases. It can be proved that the three necessary conditions for convergence to the real pdf (i.e., that $V_n \to 0$, $k_n \to \infty$, and $k_n/n \to 0$ for $n \to \infty$) are satisfied provided that the sufficient conditions $h_n = h_1/\sqrt{n}$, $\varphi(\mathbf{y}) \geq 0$ (for each $\mathbf{y} \in \mathbb{R}^d$), and $\int_{\mathbb{R}^d} \varphi(\mathbf{x}) \, d\mathbf{x} = 1$ hold. Arbitrary, crisp partitions of the feature space into regions as those often used for discretization do not satisfy the necessary conditions in the general case.

5.1.3 *How the Parzen Window Throws Light on the Debate*

The hypercubes (Equation 5.2) are conceptually similar to the popular 'intervals' (or, crisp regions) that are commonly used for discretizing continuous feature spaces (by splitting up the space into such regions, and computing the relative frequencies by counting the fraction of the data within each of them). Differences are that, in the PW approach: (*i*) the hypercubes are shifted over the definition domain according to the location of the actual vectors observed in the data sample; (*ii*) the volume they actually cover is a function of the sample, as well; (*iii*) overlapping among hypercubes centered at different locations is allowed (in fact, required); (*iv*) under-sampled regions (where the pdf has close-to-zero values) are not forced to be covered by any window functions.

In summary, insight of the PW technique and the conditions for optimal convergence to the real density function contribute to the discussion on discrete versus continuous variables, showing that:

1. Smooth, continuous-valued models (e.g., Gaussian kernels) yield more realistic estimates of density functions than crisp models (e.g., hypercubes) do;
2. Crisp partition of the feature space into intervals is a worse fit of the data than proper, usually overlapping regions are;

3. A good model of the density function can be obtained only by selecting regions that reflect the very nature of the data distribution, i.e. regions need to be densely located wherever the data are more heavily concentrated (again, crisp partition into whatever intervals is unlikely to fit);
4. In general (unless specific precautions are taken), discretization relying on any arbitrary partitioning of the continuous space does not satisfy the necessary conditions for convergence to the real pdf.

Consequently, this chapter outsprings from the conviction that continuous features may often convey relevant information, and the proposed graphical models need to be developed accordingly. The development of the hybrid random field model for continuous variables turns out to be rather straightforward, in that its definition and the corresponding theoretical/algorithmic framework remains mostly unchanged. The critical point is the selection of an adequate model for representing the conditional density functions of individual variables given their Markov blankets, and how such a model can be effectively estimated and applied within the HRF algorithms for learning and inference.

5.1.4 Chapter Outline

The issue of conditional density estimation for the continuous-valued extension of hybrid random fields is discussed in Section 5.2. Like in the traditional density estimation setup, rooted in statistics and in classic pattern recognition, three complementary philosophies may be followed: parametric, semiparametric, and nonparametric estimation. We introduced the basic idea behind parametric estimation in Section 1.2. The topic is reviewed in depth in Section 5.3, where parametric hybrid random fields are presented. They rely on the assumption that the conditional densities associated with the vertices of the graphical model have known form, and that they are uniquely determined by a vector of parameters (unknown a priori) which has to be estimated from the data. Due to their relevance to the subject, Normal distributions (Section 5.3.1) and Gaussian mixture models (Section 5.3.2) are discussed, along with the maximum-likelihood solutions of the corresponding parameter estimation problems. It is shown that while an exact solution in closed form exists for the single Gaussian density case, an iterative gradient-ascent algorithm is necessary when dealing with mixtures.

Section 5.4 presents a semiparametric conditional density estimation technique. A fundamental result, namely the 'change of variables' theorem, is stated first (Section 5.4.1). It is then used in order to derive a general learning scheme which is referred to as the nonparanormal (or nonparametric normal) approach (Section 5.4.2). The definition of nonparanormal distribution is given, basically relying on the notion that a proper mapping of the original variables exists such that the transformed feature vectors are Normally distributed. The estimation of such a mapping, and the implications on conditional independence statements entailed by the relevant precision matrix (that is, the inverse of the covariance matrix) are then investigated. Finally, it is explained how to use the mapping in order to compute (conditional) density functions over the original feature space.

The nonparametric approach to the estimation of conditional densities in hybrid random fields is presented next (Section 5.5). No assumptions on the specific form of the conditional pdfs are made in this case, and a double-kernel variant of the Parzen window technique (the Nadaraya-Watson estimator), suitable for conditional pdf estimation, is proposed. As we have said, the PW requires to define an initial bandwidth value h_1 for determining the volume of the regions involved in the estimation algorithm. Although h_1 may be fixed empirically (by means of cross-validation), a viable algorithm for automatic selection of the kernel bandwidths involved in the Nadaraya-Watson estimator is outlined in Section 5.5.2. The algorithm relies on an evaluation criterion that guides the selection process, namely the cross-validated log-likelihood. Since the computation of this criterion is heavy, it is finally explained how to tackle the complexity issue via the dual-tree recursion method, which is carefully reviewed in Section 5.5.3.

Parametric, semiparametric, and nonparametric estimation techniques in continuous hybrid random fields (and in other continuous graphical models) stand in place of what is otherwise known as 'parameter learning' in discrete graphical models. The issue of structure learning in continuous HRFs is then faced in Section 5.6, decomposing it into three sub-tasks, namely initialization of the model (Section 5.6.1), structure learning at a local level (Section 5.6.2), and structure learning at a global level (Section 5.6.3). Finally, some conclusions are drawn in Section 5.7.

5.2 Conditional Density Estimation

In continuous domains, learning probabilistic graphical models from data is much more challenging than in discrete domains. While the multinomial distribution is a generally adequate choice for estimating conditional probabilities in discrete event spaces, choosing a suitable kind of estimator for (continuous) conditional density functions requires to make a decision as to whether to assume that the form of the modeled density is known (e.g. normal), which leads to parametric techniques, or to relax the parametric assumption, which leads to nonparametric techniques [73]. The parametric assumption is often limiting, because in real-world applications the true form of the density function is rarely known *a priori*. On the other hand, nonparametric techniques only make a much weaker assumption concerning the smoothness of the density function.

In order to design suitable versions of probabilistic graphical models for continuous domains, the key challenge is to develop a technique for estimating conditional density functions. In fact, once a method for computing a (continuous) conditional distribution $p(Y \mid X_0, \ldots, X_n)$ has been defined, a continuous graphical model simply results from replacing the conditional probability tables (in Bayesian networks and hybrid random fields) or the potential function-based models by the newly defined conditional density models. Hence, the learning task which we referred to as parameter learning in the case of discrete probability distributions is now replaced by the analogous (but much more demanding) task of *conditional density estimation*.

Conditional density estimation is the problem of estimating a conditional density function from data [278]. As explained in Appendix A, the conditional density of a (continuous) random variable X given the random variable Y is defined (similarly to the discrete case) as follows:

$$p(X \mid Y) = \frac{p(X,Y)}{p(Y)} \tag{5.4}$$

Of course, the same definition holds for the conditional density of X given a random vector $\mathbf{Y}$, i.e. $p(X \mid \mathbf{Y}) = \frac{p(X,\mathbf{Y})}{p(\mathbf{Y})}$.

Given the way that a conditional density function is defined, any problem in conditional density estimation can be clearly reduced to a pair of unconditional pdf estimation problems. That is to say, if our goal is to estimate the conditional density $p(X \mid Y)$, we can address this task by estimating first the (unconditional) density functions $p(X,Y)$ and $p(Y)$, and then by computing their quotient. This approach (which is called the *quotient-shape approach* to conditional density estimation) is the one we adopt in the design of learning techniques for continuous graphical models. For completeness, we notice that some attempts have been made in the literature to devise different (and possibly more advantageous) approaches, such as the *quantile-copula approach* [83], which are currently an active investigation area in multivariate statistics. However, our treatment will be restricted to the quotient-shape approach (which is the oldest and best understood one), since a proper presentation of different conditional density estimation approaches goes beyond the scope of this book. Hence, the goal of the following sections is to present three different techniques for parametric, semiparametric, and nonparametric conditional density estimation respectively, articulated according to the quotient-shape approach.

5.3 Parametric Hybrid Random Fields

We now review two parametric density estimation techniques, which are based on Gaussian models of the data distributions. The first approach (presented in Section 5.3.1) simply assumes that the data follow a Gaussian distribution. On the other hand, in Section 5.3.2 we explain how to model data which are distributed according to a *mixture* of Gaussian distributions.

5.3.1 Normal Distributions

A random variable X is said to have a *normal* (or *Gaussian*) distribution when its probability density function is in the following, bell-shaped form:

$$p(x) = \frac{1}{\sqrt{2\pi}\,\sigma} \exp\left\{-\frac{1}{2}\left(\frac{x-\mu}{\sigma}\right)^2\right\} \tag{5.5}$$

where μ is the mean, and σ^2 is the variance of the distribution [219, 73]. The following, alternative notation may be used: $p(x) = N(x; \mu, \sigma^2)$. Whenever $\mu = 0$ and $\sigma = 1$ the distribution is said to be *standard*. The normal pdf is the most popular distribution in continuous-valued statistics, due to its relative simplicity, as well as to the fact that it describes random values that are expected to be concentrated mostly around the peak of the pdf (namely, close to its mean value μ), whilst the likelihood of observing values away from the mean decreases exponentially along the tails, according to the standard deviation σ. Normal distributions are assumed, for instance, in order to model the observational error in experimental measurements in physics and chemistry, as well as in both natural and social sciences. Figure 5.1 represents a few univariate normal pdfs having different means and variances.

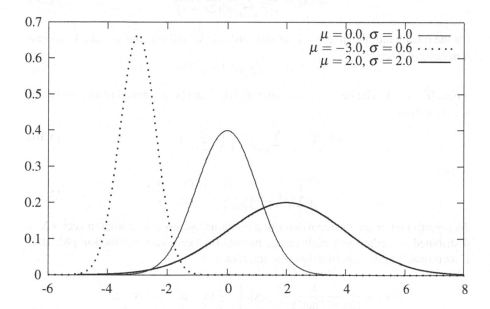

Fig. 5.1 Univariate normal pdfs with different means and variances

In an empirical perspective, if a sample is drawn from a normal distribution $N(x; \mu, \sigma^2)$, then the data are densely clustered around the mean value μ such that roughly 95% of the sample falls within the interval $|x - \mu| < 2\sigma$, and 99% of the sample is expected to be in $|x - \mu| < 3\sigma$. From a qualitative standpoint, this implies that Gaussian distributions are well suited for random variables that have a tendency to exhibit outcomes that concentrate in a limited interval centered around the mean, whilst the likelihood of observing outliers decreases exponentially with the squared distance from the mean (normal pdfs are not *heavy-tailed* distributions).

The cumulative distribution function of the standard normal distribution is defined as

$$\Phi(x) = \frac{1}{\sqrt{2\pi}} \int_{-\infty}^{x} e^{-t^2/2} \, dt$$

$$= \frac{1}{2} \left[1 + \mathrm{erf}\left(\frac{x}{\sqrt{2}} \right) \right] \tag{5.6}$$

where $\mathrm{erf}(\cdot)$ denotes the so-called *(Gauss) error function*, given by

$$\mathrm{erf}(x) = \frac{2}{\sqrt{\pi}} \int_{0}^{x} e^{-t^2} \, dt \tag{5.7}$$

The Taylor expansion for the error function is provided by Equation 5.8:

$$\mathrm{erf}(x) = \frac{2}{\sqrt{\pi}} \sum_{n=0}^{\infty} \frac{(-1)^n x^{2n+1}}{n!(2n+1)} \tag{5.8}$$

On the other hand, the inverse of the standard normal cdf (i.e. the so-called *quantile function*) is given by

$$\Phi^{-1}(x) = \sqrt{2}\, \mathrm{erf}^{-1}(2x - 1) \tag{5.9}$$

where $0 < x < 1$. The inverse error function erf^{-1} can be computed (using its Taylor expansion) as

$$\mathrm{erf}^{-1}(x) = \sum_{n=0}^{\infty} \frac{c_n}{2n+1} \left(\frac{\sqrt{\pi}}{2} x \right)^{2n+1} \tag{5.10}$$

where

$$c_n = \begin{cases} 1 & \text{if } n = 0 \\ \sum_{m=0}^{n-1} \frac{c_m c_{n-1-m}}{(m+1)(2m+1)} & \text{if } n > 0 \end{cases} \tag{5.11}$$

As regards higher-dimensional definition domains, we say that a random vector $\mathbf{X}$ is distributed according to a multivariate normal (or Gaussian) distribution [90, 104] if its probability density function is in the form:

$$p(\mathbf{x}) = \frac{1}{(2\pi)^{d/2} \det(\Sigma)^{1/2}} \exp\left\{ -\frac{1}{2}(\mathbf{x} - \boldsymbol{\mu})^{\mathsf{T}} \Sigma^{-1} (\mathbf{x} - \boldsymbol{\mu}) \right\}$$

$$= N(\mathbf{x}; \boldsymbol{\mu}, \Sigma) \tag{5.12}$$

where: $\mathbf{x} \in \mathbb{R}^d$ is a real-valued column vector; $\boldsymbol{\mu}$ is the d-dimensional *mean vector* (in column form, as well); Σ is the $d \times d$ *covariance matrix*; $\det(\Sigma)$ denotes the determinant of Σ; Σ^{-1} represents the inverse of Σ; T is the matrix transposition operator; $\frac{1}{(2\pi)^{d/2} \det(\Sigma)^{1/2}}$ is the normalization term. The quantity $(\mathbf{x} - \boldsymbol{\mu})^{\mathsf{T}} \Sigma^{-1} (\mathbf{x} - \boldsymbol{\mu})$ is known as the (quadratic) Mahalanobis distance between $\mathbf{x}$ and $\boldsymbol{\mu}$. As a consequence, regions in the definition domain for which $p(\mathbf{x})$ has a constant value (hyper-elliptical curves) are the *loci* of the vectors having constant Mahalanobis distance from the mean vector $\boldsymbol{\mu}$. The axes of these hyper-ellipses coincide with the eigenvectors of the matrix Σ. When a sample is collected, having empirical covariance matrix Σ, then the eigenvectors are sometimes referred to as the *principal components* of the sample (or, of the corresponding probability distribution).

It is immediately seen that the following, nice properties hold true for normal distributions. First of all, $E[\mathbf{X}] = \boldsymbol{\mu}$, that is, the parameter $\boldsymbol{\mu}$ of a Gaussian pdf (either univariate or multivariate) coincides with the expected value of the corresponding random variable (or random vector). Furthermore, $E[(\mathbf{x} - \boldsymbol{\mu})(\mathbf{x} - \boldsymbol{\mu})^\mathsf{T}] = \Sigma$. If $\mathbf{x} = (x_1, \ldots, x_d)$ and $\boldsymbol{\mu} = (\mu_1, \ldots, \mu_d)$ then $\mu_i = E[x_i]$ and $\sigma_{ij} = E[(x_i - \mu_i)(x_j - \mu_j)]$. Asymptotic cases worth being observed, as well: in the limit case of vanishing variance the univariate Gaussian tends to a Dirac's Delta, i.e. $\lim_{\sigma \to 0} p(\mu) = +\infty$ and $\lim_{\sigma \to 0} p(x) = 0$ for all $x \neq \mu$. On the other hand, when the variance tends to grow to infinity, the univariate normal pdf tends basically to the uniform distribution. Similar asymptotic behaviors can be observed in the multivariate case, too.

It is worth noticing that the covariance matrix is always symmetric: $\sigma_{ij} = \sigma_{ji}$. Furthermore, it is seen that the elements of Σ undergo the following properties. First of all, since the distribution of the random vector $\mathbf{X}$ can be seen as a joint distribution over its individual components, namely the random variables $X_1, \ldots, X_d$, then the i-th element σ_{ii} along the diagonal of Σ is the (univariate) variance of X_i. Again, the entry σ_{ij} of Σ designates the *covariance* of X_i and X_j, which expresses the amount of correlation between i-th and j-th components of the random vector $\mathbf{X}$. In particular, if x_i and x_j are statistically independent of each other, then $\sigma_{ij} = 0$. Consequently, if $\sigma_{ij} = 0$ for all pairs i, j such that $i \neq j$, then Σ reduces to a diagonal matrix, and its eigenvectors (principal components of the distribution) are parallel to the Cartesian coordinate axes. In this special case, the overall multivariate normal distribution reduces to a product of individual, univariate Gaussian pdfs along the different directions of the vector space, i.e., $N(\mathbf{x}; \boldsymbol{\mu}, \Sigma) = \prod_{i=1}^d N(x_i; \mu_i, \sigma_{ii})$. Due to its simplicity, this model is often assumed in real-world scenarios. In fact, although it makes a strong independence assumption which seldom fits the statistical properties of the phenomena at hand, it turns out to be much less complex and far more stable form a numerical viewpoint.

In practical circumstances, the mean and the (co-)variance of a normal distribution are not known in advance. They shall rather be inferred in some consistent mathematical framework from data collected on the field. This is a major instance of what is known as the density estimation problem in statistics [65]. The problem can be formalized as follows. Let us assume a random sample $\mathbf{D} = \{\mathbf{x}_1, \ldots, \mathbf{x}_n\}$ has been observed, where the observations (outcomes of a certain random variable, or random vector) $\mathbf{x}_i \in \mathbb{R}^d$ are independent and identically distributed (iid) according to an underlying, generic, and unknown pdf $p(\mathbf{x})$. Parametric estimation techniques make the further, strong assumption that $p(\mathbf{x})$ is in the parametric form $p(\mathbf{x}) = p(\mathbf{x} \mid \boldsymbol{\Xi})$, meaning that the form of the pdf is known (e.g., that it is Gaussian) and the very pdf is uniquely determined once specific values for the parameters $\boldsymbol{\Xi} \in \mathbb{R}^p$ (e.g., $\boldsymbol{\mu}$ and Σ) have been assigned [73, 210]. Suitable approaches to the estimation of parameters $\boldsymbol{\Xi}$ which 'optimally' fit the sample $\mathbf{D}$ are sought. The maximum-likelihood parameter estimation technique is far the most popular [73, 146], yielding particularly simple, yet robust solutions in the particular case of normal density functions. In the following we illustrate the calculations required in order to accomplish ML estimation of parameters for a generic density function $p(\mathbf{x} \mid \Xi)$ having unknown parameters Ξ, and we will later specialize the solution for Gaussian densities [73, 210].

Our aim is to exploit the information contained in $\mathbf{D}$ in order to estimate $\Xi = (\xi_1, \ldots, \xi_p)$. Given the sample $\mathbf{D} = \{\mathbf{x}_1, \ldots, \mathbf{x}_n\}$, and because of the iid assumption, we can write

$$p(\mathbf{D} \mid \Xi) = \prod_{k=1}^{n} p(\mathbf{x}_k \mid \Xi) \tag{5.13}$$

This quantity is defined to be the likelihood of the parameters given the data, and it is a function of Ξ. As such, it may be assumed as a criterion function that guides us in the choice of those parameters that result in the maximum possible value of the likelihood. More formally, the ML estimate $\widehat{\Xi}$ is defined to be the parameter vector that maximizes $p(\mathbf{D} \mid \Xi)$. The search for $\widehat{\Xi}$ can be pursued by devising analytical, necessary conditions that have to be satisfied in correspondence with the maximum of $p(\mathbf{D} \mid \Xi)$. This is best accomplished by moving to the logarithmic domain, defining the log-likelihood:

$$\begin{aligned} \ell(\Xi) &= \log p(\mathbf{D} \mid \Xi) \\ &= \sum_{k=1}^{n} \log p(\mathbf{x}_k \mid \Xi) \end{aligned} \tag{5.14}$$

Application of the logarithm (a monotonically increasing function) does not alter the *locus* in the definition domain which coincides with the maximum of the likelihood function, but is turns out to be advantageous from several viewpoints. First, it eases the calculations to a certain extent. Then, it confers numerical stability to the consequent algorithm, given that dealing with the actual numeric values of the density function (which are mostly close-to-zero over the whole definition domain) may be troublesome in practice. Finally, application of the logarithm yields particularly simple solutions in the normal case (whose core has an exponential form). The necessary conditions we were looking for can be stated as follows:

$$\nabla_{\Xi} \ell(\widehat{\Xi}) = \mathbf{0} \tag{5.15}$$

that is a system of p equations, where

$$\nabla_{\Xi} \ell(\Xi) = \sum_{k=1}^{n} \nabla_{\Xi} \log p(\mathbf{x}_k \mid \Xi) \tag{5.16}$$

is the gradient of the log-likelihood with respect to the parameter vector. In words, we search for the parameters $\widehat{\Xi}$ that result in zeros of the gradient of $\ell(\Xi)$. In order to apply Equation 5.15, calculation of the gradient is required. This strictly depends on the particular (necessarily differentiable) form of the pdf under consideration, as well as on the specific parameters Ξ whose estimation is required. Consequently, a closed-form, unique ML solution of the equation may or may not exist, according to the nature of the pdf itself. At times, several distinct solutions emerge (corresponding with different local maxima of the likelihood function). In other cases, no closed-form solution at all can be found mathematically, and approximated optimization algorithms can be used instead (we will be faced with such a scenario in Section 5.3.2).

Let us now turn our attention back to normal distributions. We start by considering the univariate case first, i.e. $p(x \mid \boldsymbol{\Xi}) = \frac{1}{\sqrt{2\pi}\sigma} \exp\left\{-\frac{1}{2}\left(\frac{x-\mu}{\sigma}\right)^2\right\}$, where $\boldsymbol{\Xi} = (\Xi_1, \Xi_2) = (\mu, \sigma^2)$. We have

$$\log p(x \mid \boldsymbol{\Xi}) = -\frac{1}{2}\log(2\pi \Xi_2) - \frac{1}{2\Xi_2}(x - \Xi_1)^2$$

and we can write

$$\nabla_{\boldsymbol{\Xi}} \log p(x \mid \boldsymbol{\Xi}) = \begin{bmatrix} \frac{1}{\Xi_2}(x - \Xi_1) \\ -\frac{1}{2\Xi_2} + \frac{(x-\Xi_1)^2}{2\Xi_2^2} \end{bmatrix}$$

In order to satisfy $\nabla_{\boldsymbol{\Xi}} \ell(\boldsymbol{\Xi}) = \mathbf{0}$, and bearing in mind that $\Xi_1 = \mu$, $\Xi_2 = \sigma^2$, we obtain the following ML solutions for the parameters $\hat{\mu}$ and $\hat{\sigma}^2$:

$$\sum_{k=1}^{n} \frac{1}{\hat{\sigma}^2}(x_k - \hat{\mu}) = 0 \qquad (5.17)$$

that yields

$$\hat{\mu} = \frac{1}{n}\sum_{k=1}^{n} x_k \qquad (5.18)$$

that is simply the sample average, as the intuition would suggest; and

$$-\sum_{k=1}^{n} \frac{1}{\hat{\sigma}^2} + \sum_{k=1}^{n} \frac{(x_k - \hat{\mu})^2}{\hat{\sigma}^4} = 0 \qquad (5.19)$$

which implies

$$\hat{\sigma}^2 = \frac{1}{n}\sum_{k=1}^{n} (x_k - \hat{\mu})^2 \qquad (5.20)$$

that, in turn, is the empirical variance of the sample with respect to its average.

In the multivariate case, i.e. when $p(\mathbf{x} \mid \boldsymbol{\Xi}) = N(\mathbf{x}; \boldsymbol{\mu}, \Sigma)$ and $\boldsymbol{\Xi} = (\boldsymbol{\mu}, \Sigma)$, similar calculations lead to the following, closed-form ML solutions for the mean vector and the covariance matrix, respectively:

$$\hat{\boldsymbol{\mu}} = \frac{1}{n}\sum_{k=1}^{n} \mathbf{x}_k \qquad (5.21)$$

and

$$\hat{\Sigma} = \frac{1}{n}\sum_{k=1}^{n} (\mathbf{x}_k - \hat{\boldsymbol{\mu}})(\mathbf{x}_k - \hat{\boldsymbol{\mu}})^{\mathsf{T}} \qquad (5.22)$$

5.3.2 Gaussian Mixture Models

Normal densities and ML estimation of their parameters are simple and popular, but they are unlikely to fit several real-world scenarios where the data are distributed in complex, unknown forms within their definition domain. Mixture models are a much more flexible and realistic approach, whose treatment benefits from the background and the calculations developed in the previous section [73, 211]. Let us consider again the sample $\mathbf{D} = \{\mathbf{x}_k \mid k = 1, \ldots, n\}$, where n vectors $\mathbf{x}_1, \ldots, \mathbf{x}_n$ are now supposed to be identically and independently drawn from the *finite mixture* density

$$p(\mathbf{x} \mid \boldsymbol{\Xi}) = \sum_{i=1}^{C} \Pi_i p_i(\mathbf{x} \mid \boldsymbol{\xi}_i) \tag{5.23}$$

where the parametric form of the component densities p_i (for $1 \leq i \leq C$) is assumed to be known, as well as the *mixing parameters* $\Pi_1, \ldots, \Pi_C$, i.e. the *a priori* probabilities of the components (the latter assumption will be relaxed shortly). Let $\boldsymbol{\Xi} = (\boldsymbol{\xi}_1, \ldots, \boldsymbol{\xi}_C)$ be the vector of all parameters associated with each component density (from now on, $\boldsymbol{\xi}_i$ denotes the unknown parameters of i-th component density). We make two further assumptions: (1) i-th parameter vector $\boldsymbol{\xi}_i$ is functionally independent of j-th parameter vector $\boldsymbol{\xi}_j$ if $i \neq j$; (2) the mixture is *identifiable*, i.e. it is such that, if $\boldsymbol{\Xi} \neq \boldsymbol{\Xi}^*$, then $\exists \mathbf{x} \, p(\mathbf{x} \mid \boldsymbol{\Xi}) \neq p(\mathbf{x} \mid \boldsymbol{\Xi}^*)$, where $\mathbf{x} \in \mathbb{R}^d$. In the present setup, we want to use the data sample to estimate the parameters $\boldsymbol{\Xi}$.

Assuming the data are independently drawn from $p(\mathbf{x} \mid \boldsymbol{\Xi})$, the likelihood of a certain choice of parameters $\boldsymbol{\Xi}$ given the observed data $\mathbf{D}$ can be written as:

$$p(\mathbf{D} \mid \boldsymbol{\Xi}) = \prod_{j=1}^{n} p(\mathbf{x}_j \mid \boldsymbol{\Xi}). \tag{5.24}$$

As in the previous section, ML estimation techniques search for the parameters $\boldsymbol{\Xi}$ that maximize the value indicated in Equation 5.24 or, equivalently, the log-likelihood [73, 299]:

$$\begin{aligned} \ell(\boldsymbol{\Xi}) &= \log p(\mathbf{D} \mid \boldsymbol{\Xi}) \\ &= \sum_{j=1}^{n} \log p(\mathbf{x}_j \mid \boldsymbol{\Xi}) \end{aligned} \tag{5.25}$$

Let us now exploit the functional independence assumption. If a certain parameter vector $\widehat{\boldsymbol{\Xi}}$ maximizes $\ell(\widehat{\boldsymbol{\Xi}})$, then it has to satisfy the following necessary condition:

$$\nabla_{\boldsymbol{\xi}_i} \ell(\widehat{\boldsymbol{\Xi}}) = \mathbf{0} \tag{5.26}$$

for $1 \leq i \leq C$, where the operator $\nabla_{\boldsymbol{\xi}_i}$ denotes the gradient vector computed with respect to the parameters $\widehat{\boldsymbol{\xi}}_i$, as usual, and $\mathbf{0}$ is the vector with all components equal to zero. In other words, we are looking for the zeros of the gradient of the log-likelihood, as we did in the previous Section. Substituting Equation 5.23 into Equation 5.25 and the latter, in turn, into Equation 5.26, we can write:

$$\nabla_{\hat{\boldsymbol{\xi}}_i} \ell(\widehat{\boldsymbol{\Xi}}) = \sum_{j=1}^{n} \nabla_{\hat{\boldsymbol{\xi}}_i} \log p(\mathbf{x}_j \mid \widehat{\boldsymbol{\Xi}})$$

$$= \sum_{j=1}^{n} \nabla_{\hat{\boldsymbol{\xi}}_i} \log \sum_{i=1}^{C} \Pi_i p_i(\mathbf{x}_j \mid \hat{\boldsymbol{\xi}}_i) \qquad (5.27)$$

$$= \sum_{j=1}^{n} \frac{1}{p(\mathbf{x}_j \mid \widehat{\boldsymbol{\Xi}})} \nabla_{\hat{\boldsymbol{\xi}}_i} \Pi_i p_i(\mathbf{x}_j \mid \hat{\boldsymbol{\xi}}_i)$$

$$= \mathbf{0}$$

Using Bayes' Theorem we have:

$$P(i \mid \mathbf{x}_j, \hat{\boldsymbol{\xi}}_i) = \frac{\Pi_i p_i(\mathbf{x}_j \mid \hat{\boldsymbol{\xi}}_i)}{p(\mathbf{x}_j \mid \widehat{\boldsymbol{\Xi}})} \qquad (5.28)$$

where $P(i \mid \mathbf{x}_j, \hat{\boldsymbol{\xi}}_i)$ is the *a posteriori* probability of i-th component given the observation $\mathbf{x}_j$ and the parameters $\hat{\boldsymbol{\xi}}_i$. Equation 5.27 can thus be rewritten as:

$$\nabla_{\hat{\boldsymbol{\xi}}_i} \ell(\widehat{\boldsymbol{\Xi}}) = \sum_{j=1}^{n} \frac{P(i \mid \mathbf{x}_j, \hat{\boldsymbol{\xi}}_i)}{\Pi_i p_i(\mathbf{x}_j \mid \hat{\boldsymbol{\xi}}_i)} \nabla_{\hat{\boldsymbol{\xi}}_i} \Pi_i p_i(\mathbf{x}_j \mid \hat{\boldsymbol{\xi}}_i)$$

$$= \sum_{j=1}^{n} P(i \mid \mathbf{x}_j, \hat{\boldsymbol{\xi}}_i) \nabla_{\hat{\boldsymbol{\xi}}_i} \log \Pi_i p_i(\mathbf{x}_j \mid \hat{\boldsymbol{\xi}}_i) \qquad (5.29)$$

$$= \mathbf{0}$$

In the following, we will concentrate our attention on the case in which the component densities of the mixture are multivariate normal distributions. In this case the parametric pdf under consideration is called a Gaussian mixture model (GMM). GMMs are of particular interest since, apart from their relative simplicity, they are 'universal' models, in that a GMM exists that can approximate any given, continuous pdf to an arbitrary degree of precision. It goes without saying that this property does not tell us how many components are needed in order to match the actual density function underlying a given data sample, neither does it tell us which parameters shall be assigned to the components in order to reach the optimal fit. Again, the following ML estimation algorithm is not guaranteed to find the parameters that yield the absolute maximum of the likelihood function.

Formally, a GMM is a mixture density whose components have the following form:

$$p_i(\mathbf{x}_j \mid \boldsymbol{\xi}_i) = N(\mathbf{x}_j; \boldsymbol{\mu}_i, \Sigma_i) \qquad (5.30)$$

according to the definition of multivariate normal pdf given by Equation 5.12. The parameters to be estimated for the i-th component density are its mean vector and its covariance matrix, that is to say:

$$\boldsymbol{\xi}_i = (\boldsymbol{\mu}_i, \Sigma_i) \tag{5.31}$$

Substituting Equation 5.12 into Equation 5.30 and the latter, in turn, into Equation 5.29, we can write the gradient of the log-likelihood for the case of normal component densities as:

$$\nabla_{\boldsymbol{\xi}_i} \ell(\boldsymbol{\Xi}) =$$
$$= \sum_{j=1}^{n} P(i \mid \mathbf{x}_j, \boldsymbol{\xi}_i) \nabla_{\boldsymbol{\xi}_i} \left\{ \log \Pi_i (2\pi)^{-\frac{d}{2}} \det(\Sigma_i)^{-\frac{1}{2}} - \frac{1}{2} (\mathbf{x}_j - \boldsymbol{\mu}_i)^{\mathsf{T}} \Sigma_i^{-1} (\mathbf{x}_j - \boldsymbol{\mu}_i) \right\} \tag{5.32}$$

Suppose first that the only unknown parameters to be estimated are the mean vectors of the Gaussian distributions, e.g. the covariances are assumed to be known in advance (also, this assumption will be dropped shortly). There are practical situations, for example in data clustering [73, 211], in which the estimation of the means is sufficient, but the extension to the more general case of unknown covariances will turn out to be simple, as well. By setting $\boldsymbol{\Xi} = (\boldsymbol{\mu}_1, \dots, \boldsymbol{\mu}_C)$, Equation 5.32 reduces to:

$$\nabla_{\boldsymbol{\xi}_i} \ell(\boldsymbol{\Xi}) = \sum_{j=1}^{n} \left\{ P(i \mid \mathbf{x}_j, \boldsymbol{\mu}_i) \nabla_{\boldsymbol{\mu}_i} \left[-\frac{1}{2} (\mathbf{x}_j - \boldsymbol{\mu}_i)^{\mathsf{T}} \Sigma_i^{-1} (\mathbf{x}_j - \boldsymbol{\mu}_i) \right] \right\}$$
$$= \sum_{j=1}^{n} P(i \mid \mathbf{x}_j, \boldsymbol{\mu}_i) \Sigma_i^{-1} (\mathbf{x}_j - \boldsymbol{\mu}_i). \tag{5.33}$$

Again, we are looking for the parameters $\widehat{\boldsymbol{\Xi}} = (\hat{\boldsymbol{\mu}}_1, \dots, \hat{\boldsymbol{\mu}}_C)$ that maximize the log-likelihood, i.e., that correspond to a zero of its gradient. From Equation 5.33, setting $\nabla_{\boldsymbol{\xi}_i} \ell(\widehat{\boldsymbol{\Xi}}) = \mathbf{0}$ allows us to write:

$$\sum_{j=1}^{n} P(i \mid \mathbf{x}_j, \hat{\boldsymbol{\mu}}_i) \mathbf{x}_j = \sum_{j=1}^{n} P(i \mid \mathbf{x}_j, \hat{\boldsymbol{\mu}}_i) \hat{\boldsymbol{\mu}}_i \tag{5.34}$$

This leads to the following central equation:

$$\hat{\boldsymbol{\mu}}_i = \frac{\sum_{j=1}^{n} P(i \mid \mathbf{x}_j, \hat{\boldsymbol{\mu}}_i) \mathbf{x}_j}{\sum_{j=1}^{n} P(i \mid \mathbf{x}_j, \hat{\boldsymbol{\mu}}_i)} \tag{5.35}$$

which shows that the ML estimate for the i-th mean vector is a weighted average over the sample (i.e., the training data), where each observation $\mathbf{x}_j$ gives a contribution that is proportional to the *estimated* probability of i-th component given the observation itself. Equation 5.35 can not be explicitly solved, but it can be put in a quite interesting and practical iterative form by making explicit the dependence of the current estimate on the number t of iterative steps that have already been accomplished:

$$\hat{\boldsymbol{\mu}}_i^{t+1} = \frac{\sum_{j=1}^n P(i \mid \mathbf{x}_j, \hat{\boldsymbol{\mu}}_i^t) \mathbf{x}_j}{\sum_{j=1}^n P(i \mid \mathbf{x}_j, \hat{\boldsymbol{\mu}}_i^t)} \tag{5.36}$$

where $\hat{\boldsymbol{\mu}}_i^t$ denotes the estimate obtained at t-th iterative step, and the corresponding value is actually used to compute the new estimate at $(t+1)$-th step.

If also Σ_j and Π_j are considered to be parameters to be estimated from the data, similar calculations [73] lead to the following ML re-estimation formulas:

$$\hat{\Pi}_j = \frac{1}{n} \sum_{k=1}^n P(j \mid \mathbf{x}_k, \hat{\Xi}) \tag{5.37}$$

and

$$\hat{\Sigma}_j = \frac{\sum_{k=1}^n P(j \mid \mathbf{x}_k, \hat{\Xi})(\mathbf{x}_k - \hat{\boldsymbol{\mu}}_j)(\mathbf{x}_k - \hat{\boldsymbol{\mu}}_j)^\mathsf{T}}{\sum_{k=1}^n P(j \mid \mathbf{x}_k, \hat{\Xi})} \tag{5.38}$$

which can be applied in an iterative fashion, as we did through Equation 5.36 for the mean vectors. Note that these iterative ML algorithms require to start up from a (somewhat arbitrary) initial choice for the parameters under consideration. Random initialization can be assumed, but a much more meaningful initialization scheme will be introduced shortly.

In passing, let us consider the fact that $P(i \mid \mathbf{x}_j, \hat{\boldsymbol{\mu}}_i^t)$ is large when the Mahalanobis distance between $\mathbf{x}_j$ and $\hat{\boldsymbol{\mu}}_i^t$ is small. Then, it appears to be reasonable to estimate an approximation of $P(i \mid \mathbf{x}_j, \hat{\boldsymbol{\mu}}_i^t)$ in the following way [73, 299]:

$$P(i \mid \mathbf{x}_j, \hat{\boldsymbol{\mu}}_i^t) \approx \begin{cases} 1 & \text{if } dist(\mathbf{x}_j, \hat{\boldsymbol{\mu}}_i^t) = \min_{l=1,\dots,C} dist(\mathbf{x}_j, \hat{\boldsymbol{\mu}}_l^t) \\ 0 & \text{otherwise} \end{cases} \tag{5.39}$$

for a given distance measure $dist(\cdot, \cdot)$. Mahalanobis distance should be used, but Euclidean distance is usually an effective choice, so that Equation 5.36 reduces to:

$$\hat{\boldsymbol{\mu}}_i^{t+1} = \frac{1}{n_i^t} \sum_{k=1}^{n_i^t} \mathbf{x}_k^i \tag{5.40}$$

where n_i^t is the estimated number of training observations drawn from i-th component at step t (i.e. the number of vectors for which $P(i \mid \mathbf{x}_j, \hat{\boldsymbol{\mu}}_i^t) = 1$ according to Equation 5.39) and $\mathbf{x}_k^i$ is the k-th of these observations (for $1 \leq k \leq n_i^t$). Equation 5.40 is known as the *k-means* clustering algorithm [73, 151]. The latter can thus be interpreted as an approximated ML solution to the estimate of mean vectors of a mixture of Normal densities under the assumptions we made above. Although clustering techniques are beyond the scope of this book, Equation 5.40 is a simple, yet effective initialization technique for the parameter vectors that have to undergo the overall, iterative ML re-estimation algorithm, namely Equations 5.36–5.38. These algorithms, as well as the k-means clustering, are particularly relevant instances of the broad expectation-maximization (EM) algorithm [62]. For this reason, they are sometimes referred to as the EM parameter estimation technique for GMMs.

5.4 Semiparametric Hybrid Random Fields

This section presents a semiparametric method for estimating (conditional) density functions. This technique (which is referred to as *nonparanormal*) was introduced in [197] for learning the structure of (sparse) undirected graphs. However, one of the contributions of this book is to show how the nonparanormal approach can be further developed so as to come up with a general-purpose density estimation method. The issue of nonparanormal (conditional) density estimation is addressed in Section 5.4.2. Before that, Section 5.4.1 reviews an important result in multivariate statistics, which is needed in order to establish the basic properties of the nonparanormal technique.

5.4.1 Change of Variables

We now state an important lemma from multivariate calculus, which is commonly known as the *change of variables* theorem [161]:

Lemma 5.1. *Consider two random vectors* $\mathbf{X}$ *and* $\mathbf{Y}$*, with domains* $\mathcal{X} \subseteq \mathbb{R}^d$ *and* $\mathcal{Y} \subseteq \mathbb{R}^d$ *respectively. Suppose that* $f : \mathcal{X} \to \mathcal{Y}$ *is a one-to-one, differentiable function from* $\mathcal{X}$ *onto* $\mathcal{Y}$*. Then, if* $\mathbf{X}$ *and* $\mathbf{Y}$ *are distributed according to density functions* $p_{\mathbf{X}}(\mathbf{x})$ *and* $p_{\mathbf{Y}}(\mathbf{y})$ *respectively, it follows that*

$$p_{\mathbf{Y}}(\mathbf{y}) = p_{\mathbf{X}}(\mathbf{x}) \left| \det \frac{\partial \mathbf{x}}{\partial \mathbf{y}} \right| \qquad (5.41)$$

where $f^{-1} : \mathcal{Y} \to \mathcal{X}$ *is the inverse of* f*,* $\mathbf{x} = f^{-1}(\mathbf{y})$*, and* $\frac{\partial \mathbf{x}}{\partial \mathbf{y}}$ *denotes the Jacobian matrix of* f^{-1}*.*

Proof. See e.g. [161]. □

5.4.2 The Nonparanormal

The nonparanormal (or *nonparametric normal*) approach is a recently introduced technique for estimating the structure of undirected graphs, based on a Gaussian model, without making any parametric assumption concerning the form of the modeled density [197]. Although the previous statement may seem paradoxical, the idea underlying the nonparanormal approach is to map a set of data points (which are not known to be normally distributed) onto a set of data points that can be assumed to follow a normal distribution. Once the density of the normal sample has been estimated using a standard Gaussian model, the density of the points in the original space can then be recovered by applying the change of variables theorem.

First of all, let us define the concept of nonparanormal density:

Definition 5.1. A random vector $\mathbf{X} = (X_1, \ldots, X_d)$ with mean $\boldsymbol{\mu}$ is said to be nonparanormally distributed if there exists a function f such that:

1. $f(\mathbf{X}) = (f_1(X_1), \ldots, f_d(X_d))$, where $f_i(X_i)$ is one-to-one and differentiable (for $1 \leq i \leq d$);
2. the random vector $\mathbf{Y} = f(\mathbf{X})$ is distributed normally with mean $\boldsymbol{\mu}$ and covariance matrix Σ.

Given Definition 5.1, we can prove the following theorem:

Theorem 5.1. *If the distribution of a random vector $\mathbf{X} = X_1, \ldots, X_d$ is nonparanormal with mapping $f(\mathbf{X}) = (f_1(X_1), \ldots, f_d(X_d))$, then the density of X is given by*

$$p_{\mathbf{X}}(\mathbf{x}) = \frac{1}{(2\pi)^{d/2} \det(\Sigma)^{1/2}} \exp\left\{-\frac{1}{2}(\mathbf{y} - \boldsymbol{\mu})^{\mathsf{T}} \Sigma^{-1}(\mathbf{y} - \boldsymbol{\mu})\right\} \prod_{i=1}^{d} \left|\frac{\mathrm{d}}{\mathrm{d}x_i} f_i(x_i)\right| \quad (5.42)$$

where $\mathbf{y} = f(\mathbf{x})$, $\boldsymbol{\mu}$ is the mean vector of both $\mathbf{X}$ and $\mathbf{Y}$, and Σ is the covariance matrix of $\mathbf{Y}$.

Proof. Let $p_{\mathbf{Y}}$ denote the (normal) density function of $\mathbf{Y}$. That is,

$$p_{\mathbf{Y}}(\mathbf{y}) = \frac{1}{(2\pi)^{d/2} \det(\Sigma)^{1/2}} \exp\left\{-\frac{1}{2}(\mathbf{y} - \boldsymbol{\mu})^{\mathsf{T}} \Sigma^{-1}(\mathbf{y} - \boldsymbol{\mu})\right\} \quad (5.43)$$

Then, Lemma 5.1 implies that

$$p_{\mathbf{X}}(\mathbf{x}) = \frac{1}{(2\pi)^{d/2} \det(\Sigma)^{1/2}} \exp\left\{-\frac{1}{2}(\mathbf{y} - \boldsymbol{\mu})^{\mathsf{T}} \Sigma^{-1}(\mathbf{y} - \boldsymbol{\mu})\right\} \left|\det\frac{\partial \mathbf{y}}{\partial \mathbf{x}}\right| \quad (5.44)$$

Since the value of each f_i only depends on x_i, the Jacobian matrix $\frac{\partial \mathbf{y}}{\partial \mathbf{x}}$ is diagonal. Therefore, the absolute value of the Jacobian determinant is given by

$$\left|\det\frac{\partial \mathbf{y}}{\partial \mathbf{x}}\right| = \prod_{i=1}^{d} \left|\frac{\mathrm{d}}{\mathrm{d}x_i} f_i(x_i)\right| \quad (5.45)$$

$\square$

Based on Theorem 5.1, the crucial problem for the nonparanormal approach is how to estimate the functions $f_1(X_1), \ldots, f_d(X_d)$. The technique developed in [197] prescribes to estimate the value of each f_i as

$$\hat{f}_i(x) = \hat{\mu}_i + \hat{\sigma}_i \hat{h}_i(x) \quad (5.46)$$

where $\hat{\mu}_i$ and $\hat{\sigma}_i$ are the sample mean and standard deviation of variable X_i, and $\hat{h}_i(x)$ is defined as follows:

$$\hat{h}_i(x) = \Phi^{-1}(\widehat{F}_i(x)) \quad (5.47)$$

In Equation 5.47, Φ^{-1} is the inverse of the standard normal cdf Φ, whereas $\widehat{F}_i$ is the so-called *truncated* estimator of the empirical cdf of X_i [69], denoted by F_i^E. Let n be the number of data points and δ_n be a truncation parameter. The truncated estimator of F_i^E is then defined as

$$
\widehat{F}_i(x) = \begin{cases} \delta_n & \text{if } F_i^E(x) < \delta_n \\ F_i^E(x) & \text{if } \delta_n \leq F_i^E(x) \leq 1 - \delta_n \\ 1 - \delta_n & \text{if } 1 - \delta_n < F_i^E(x) \end{cases} \tag{5.48}
$$

The suggested setting for the truncation parameter is given by

$$
\delta_n = \frac{1}{4n^{1/4}\sqrt{\pi \log n}} \tag{5.49}
$$

It is reported in [197] that the choice specified in Equation 5.49 results in a generally satisfying behavior of the nonparanormal estimator, especially in the high-dimensional setting.

As defined in Equation 5.48, the truncated estimator $\widehat{F}_i(x)$ is discontinuous. This prevents us from computing the derivatives contained in Equation 5.42. In other words, the approach described thus far is not yet suitable as a thorough density estimation technique. On the other hand, it is sufficient instead for estimating the structure of the undirected graph underlying the density $p_\mathbf{Y}$, as this structure is conveyed by the precision matrix $\Omega = \Sigma^{-1}$ [197]. In fact, the matrix Ω is such that, if $\Omega_{ij} = 0$, then the nodes X_i and X_j will not be adjacent in the graph of a Markov random field representing $p_\mathbf{Y}$.[1] Now, one key result proved in [197] is the following lemma:

Lemma 5.2. *If the distribution of a random vector* $\mathbf{X} = (X_1, \ldots, X_d)$ *is nonparanormal with mapping* $\mathbf{Y} = f(\mathbf{X})$, *and* Ω *is the precision matrix of* $\mathbf{Y}$, *then* X_i *is conditionally independent of* X_j *given the set* $\mathcal{S} = \{X_k : 1 \leq k \leq d \wedge k \neq i, j\}$ *if and only if* $\Omega_{ij} = 0$.

Proof. See [197].   □

In other words, Lemma 5.2 states that, if $\mathbf{X}$ is nonparanormal with mapping $\mathbf{Y} = f(\mathbf{X})$, then X_i is independent of X_j given a subset $\mathcal{S}_\mathbf{X}$ of $\{X_1, \ldots, X_d\}$ if and only if Y_i is independent of Y_j given the set $\mathcal{S}_\mathbf{Y} = \{Y_k : X_k \in \mathcal{S}_\mathbf{X}\}$. Hence, the strategy explored in [197] consists in exploiting the results given above for learning the structure of Markov random fields. To this aim, the guiding idea is that the precision matrix Ω fixes not only the graph of a Markov network for $p_\mathbf{Y}$, but also the graph of a Markov random field for $p_\mathbf{X}$, as entailed by Lemma 5.2. However, since our interest lies in exploiting the nonparanormal approach for the sake of (conditional)

---

[1] In general, the precision matrix Ω of a (normally distributed) random vector $\mathbf{X}$ is such that, if $\Omega_{ij} = 0$, then X_i is independent of X_j given the set $\mathcal{S} = \{X_k : 1 \leq k \leq d \wedge k \neq i, j\}$, that is $p(X_i \mid X_j, \mathcal{S}) = p(X_i \mid \mathcal{S})$ [189].

density estimation, we now move beyond the strategy presented above, trying to fit the approach to our overall goal.

A differentiable approximation $F_i^*(x)$ of $F_i^E(x)$ can be obtained by using the sigmoid function:

$$F_i^*(x) = \frac{1}{n} \sum_{j=1}^{n} \frac{1}{1 + \exp\left(-\frac{x-x_j}{h}\right)} \tag{5.50}$$

where h is a parameter controlling the smoothness of the sigmoid. Given the approximation of Equation 5.50, we replace $\hat{h}_i(x)$ by $\hat{h}_i^*(x)$:

$$\hat{h}_i^*(x) = \Phi^{-1}\left(F_i^*(x)\right) \tag{5.51}$$

An approximate value of the derivatives referred to in Equation 5.42 can then be estimated as follows:

$$
\begin{aligned}
\frac{d}{dx}\hat{f}_i(x) &= \frac{d}{dx}\left(\hat{\mu}_i + \hat{\sigma}_i\hat{h}_i^*(x)\right) \\
&= \hat{\sigma}_i \frac{d}{dx}\hat{h}_i^*(x) \\
&= \hat{\sigma}_i \frac{d}{dx}\Phi^{-1}\left(F_i^*(x)\right) \\
&= \hat{\sigma}_i \frac{d}{dF_i^*(x)}\Phi^{-1}\left(F_i^*(x)\right)\frac{d}{dx}F_i^*(x)
\end{aligned}
\tag{5.52}
$$

First, we need to derive $\Phi^{-1}\left(F_i^*(x)\right)$ with respect to $F_i^*(x)$:

$$
\begin{aligned}
\frac{d}{dF_i^*(x)}\Phi^{-1}\left(F_i^*(x)\right) &= \frac{1}{\frac{d}{d\Phi^{-1}(F_i^*(x))}\Phi\left(\Phi^{-1}(F_i^*(x))\right)} \\
&= \frac{\sqrt{2\pi}}{\exp\left(-\frac{1}{2}\left(\Phi^{-1}(F_i^*(x))\right)^2\right)} \\
&= \frac{\sqrt{2\pi}}{\exp\left(-\frac{1}{2}\left(\sqrt{2}\,\mathrm{erf}^{-1}(2F_i^*(x)-1)\right)^2\right)} \\
&= \frac{\sqrt{2\pi}}{\exp\left(-\mathrm{erf}^{-1}\left(2F_i^*(x)-1\right)^2\right)}
\end{aligned}
\tag{5.53}
$$

We now derive $F_i^*(x)$ with respect to x:

$$
\begin{aligned}
\frac{d}{dx}F_i^*(x) &= \frac{d}{dx}\left(\frac{1}{n}\sum_{j=1}^{n}\frac{1}{1+\exp(-\frac{x-x_j}{h})}\right)\\
&= \frac{1}{n}\sum_{j=1}^{n}\frac{d}{dx}\frac{1}{1+\exp\left(-\frac{x-x_j}{h}\right)}\\
&= -\frac{1}{n}\sum_{j=1}^{n}\frac{\frac{d}{dx}\exp\left(-\frac{x-x_j}{h}\right)}{\left(1+\exp\left(-\frac{x-x_j}{h}\right)\right)^2}\\
&= \frac{1}{nh}\sum_{j=1}^{n}\frac{\exp\left(-\frac{x-x_j}{h}\right)}{\left(1+\exp\left(-\frac{x-x_j}{h}\right)\right)^2}
\end{aligned}
\tag{5.54}
$$

Given Equations 5.53–5.54, we can state the following conclusion:

$$
\frac{d}{dx}\hat{f}_i(x) = \frac{\hat{\sigma}_i\sqrt{2\pi}}{nh\exp\left(-\mathrm{erf}^{-1}\left(2F_i^*(x)-1\right)^2\right)}\sum_{j=1}^{n}\frac{\exp\left(-\frac{x-x_j}{h}\right)}{\left(1+\exp\left(-\frac{x-x_j}{h}\right)\right)^2}
\tag{5.55}
$$

If the value given in Equation 5.55 is substituted into Equation 5.42, the resulting model can be straightforwardly used as a conditional density estimator based on the quotient-shape approach.

5.5 Nonparametric Hybrid Random Fields

In directed and undirected graphical models, nonparametric conditional density estimators (based on Parzen windows [234, 73]) are used for the first time in [136, 137]. With respect to these models, in continuous HRFs we not only exploit double-kernel estimators (instead of single-kernel Parzen windows), but we also automate the task of bandwidth selection.

A nonparametric technique for learning the structure of continuous BNs is also developed in [206]. However, that method is only aimed at inferring the conditional independencies from data, rather than at learning the overall density function. A semiparametric technique for learning undirected graphs, leading to nonparanormal Markov random fields, is proposed in [197], as discussed in Section 5.4.2. In this case, after mapping the original data points onto a set of normally distributed points, the graph is estimated from the transformed dataset using the graphical lasso technique [98], which is both computationally efficient and theoretically sound for Gaussian distributions [253]. The idea of mapping the original dataset into a feature space where data are assumed to be normally distributed is also exploited in [10].

In Section 5.5.1, we describe a kernel-based conditional density estimator, which is commonly known as the *Nadaraya-Watson estimator*. Section 5.5.2 presents instead a method for tuning the so-called 'bandwidth' parameters of the kernel-based estimator. Finally, in Section 5.5.3 we review the *dual-tree recursion* approach, which is a technique for accelerating kernel-based computations, so as to make density estimation scale better with the size of the training dataset.

5.5.1 Kernel-Based Conditional Density Estimation

In order to estimate the conditional density $p(y \mid x)$, where y is the value of a random variable Y and x is the value of a random vector X, we use the Nadaraya-Watson (NW) estimator [223, 311, 263]. Suppose we are given a dataset $D = \{(x_1, y_1), \ldots, (x_n, y_n)\}$. Then, the estimator takes the following form:

$$\hat{p}(y \mid x) = \frac{\sum_{i=1}^{n} K_{h_1}(y - y_i) K_{h_2}(x - x_i)}{\sum_{i=1}^{n} K_{h_2}(x - x_i)} \tag{5.56}$$

In Equation 5.56, each function K_h is defined as follows:

$$K_h(u) = \frac{1}{h^d} K\left(\frac{\|u\|}{h}\right) \tag{5.57}$$

where K is a kernel function, h is the bandwidth (or window width), i.e. a parameter determinining the width of the kernel function, and d is the dimensionality of u. Our choice for K is the Epanechnikov kernel [80]:

$$K(x) = \frac{3}{4}(1 - x^2)\mathbf{1}_{\{|x| \leq 1\}} \tag{5.58}$$

where

$$\mathbf{1}_{\{|x| \leq 1\}} = \begin{cases} 1 & \text{if } |x| \leq 1 \\ 0 & \text{otherwise} \end{cases} \tag{5.59}$$

We use the Epanechnikov kernel not only because it is known to be asymptotically optimal, but also because it offers a significant computational advantage (at least in the presence of large datasets) with respect to other optimal functions such as the Gaussian kernel [278].

5.5.2 Bandwidth Selection

In order for the NW estimator to deliver accurate predictions, it is crucial to choose suitable values for the bandwidths h_1 and h_2. Our strategy for dealing with this task is based on the idea of finding the bandwidth values that maximize the cross-validated log-likelihood (CVLL) of the estimator given the dataset D [278, 138]. CVLL can be defined as follows:

$$CVLL(h_1, h_2) = \frac{1}{n} \sum_{i=1}^{n} \log\left(\hat{p}^{-i}(y_i \mid x_i)\hat{p}^{-i}(x_i)\right) \tag{5.60}$$

where

$$\hat{p}^{-i}(y_i \mid \mathbf{x}_i) = \frac{\sum_{j \neq i} K_{h_1}(y_i - y_j) K_{h_2}(\mathbf{x}_i - \mathbf{x}_j)}{\sum_{j \neq i} K_{h_2}(\mathbf{x}_i - \mathbf{x}_j)} \tag{5.61}$$

and

$$\hat{p}^{-i}(\mathbf{x}_i) = \frac{1}{n-1} \sum_{j \neq i} K_{h_2}(\mathbf{x}_i - \mathbf{x}_j) \tag{5.62}$$

Simplifying Equation 5.60, we get:

$$CVLL(h_1, h_2) = \left(\frac{1}{n} \sum_{i=1}^{n} \log \sum_{j \neq i} K_{h_1}(y_i - y_j) K_{h_2}(\mathbf{x}_i - \mathbf{x}_j) \right) - \log(n-1) \tag{5.63}$$

The algorithm that we develop in order to maximize $CVLL(h_1, h_2)$ performs a double dichotomic search in a space of possible bandwidth pairs. Two ranges of values $(0, h_{max_1})$ and $(0, h_{max_2})$ are simultaneously explored by evaluating subregions of the intervals (according to the CVLL metric) and then narrowing down the search to smaller intervals in an iterative way. An iteration of the algorithm begins by splitting each interval $(0, h_{max_i})$ in two (equally large) regions $\mathcal{H}_{i_1}$ and $\mathcal{H}_{i_2}$. Then, each pair $(\mathcal{H}_{1_i}, \mathcal{H}_{2_j})$ such that $i, j \in \{1, 2\}$ is evaluated by choosing the median of each region as the value of the corresponding bandwidth. Finally, the pair of regions that maximizes the CVLL is selected as pair of (narrower) intervals for the following iteration. The algorithm returns the highest-scoring pair (h_1, h_2) found during the search (see Algorithm 5.1).

5.5.3 Dual-Tree Recursion

Clearly, the complexity of computing the CVLL function is quadratic in the number of data points. This can be a serious limitation when dealing with very large datasets. However, a promising way of overcoming this issue is proposed in [138, 139] based on dual-tree recursion [120].

Dual-tree recursion is a technique for exploiting a partitioning of the feature space, as achieved via a kd-tree [23], in order to speed up kernel-based computations [120]. The basic idea is the following. Suppose that the feature space X has been partitioned into a set of regions in such a way that, if $\mathbf{D}$ is the dataset at hand, then for each region there is a corresponding subsample $\mathcal{R}_i$ such that $\mathcal{R}_i \subseteq \mathbf{D}$. Now, if we need to evaluate a certain function $f(\mathbf{x}, \mathbf{y})$ for each possible pair of points in $\mathbf{D}$, then it may be the case that, for some particular subsamples $\mathcal{R}_i$ and $\mathcal{R}_j$, the value of the function is approximately constant for pairs $(\mathbf{x}, \mathbf{y})$ such that $\mathbf{x} \in \mathcal{R}_i$ and $\mathbf{y} \in \mathcal{R}_j$. In this case, it will not be necessary to compute the value of $f(\mathbf{x}, \mathbf{y})$ for any possible choice of $\mathbf{x}$ and $\mathbf{y}$. Instead, the value computed for an arbitrary choice of $(\mathbf{x}, \mathbf{y})$ from $\mathcal{R}_i \times \mathcal{R}_j$ will be assumed to hold (up to a certain error bound) for any pair $(\mathbf{x}^*, \mathbf{y}^*)$ such that $(\mathbf{x}^*, \mathbf{y}^*) \in \mathcal{R}_i \times \mathcal{R}_j$.

Algorithm 5.1 `DDSSelectBandwidths`: Bandwidth selection by double dichotomic search

---

Input: Dataset $\mathbf{D}$; limit points h_{max_1}, h_{max_2}; number s of iterations.
Output: Bandwidth pair (h_1, h_2).

```
DDSSelectBandwidths(D, h_max1, h_max2, s):
 1.    maxScore = -∞
 2.    for(i = 1 to 2)
 3.        min_i = 0
 4.        max_i = h_max_i
 5.        median_i = ½max_i
 6.    for(i = 1 to s)
 7.        for(j = 1 to 2)
 8.            ε_j = ¼(max_j - min_j)
 9.            h_j1 = median_j - ε_j
10.            h_j2 = median_j + ε_j
11.        (k, k') = argmax_{k,k'∈{1,2}} CVLL(h_1k, h_2k')
12.        median_1 = h_1k
13.        median_2 = h_2k'
14.        for(j = 1 to 2)
15.            min_j = median_j - ε_j
16.            max_j = median_j + ε_j
17.        if(CVLL(h_1k, h_2k') > maxScore)
18.            maxScore = CVLL(h_1k, h_2k')
19.            h_1 = h_1k
20.            h_2 = h_2k'
21.    return (h_1, h_2)
```

Consider the form of the CVLL function defined in Equation 5.63. If the feature space is partitioned into m regions $\mathcal{R}_1, \ldots, \mathcal{R}_m$, then Equation 5.63 can be rewritten as follows:

$$CVLL(h_1, h_2) = \left(\frac{1}{n} \sum_{k=1}^{m} \sum_{\mathbf{z}_i \in \mathcal{R}_k} \log \sum_{l=1}^{m} \sum_{\mathbf{z}_j \in \mathcal{R}_l^{-i}} K^*(\mathbf{z}_i, \mathbf{z}_j) \right) - \log(n-1) \qquad (5.64)$$

where $\mathbf{z}_i = (\mathbf{x}_i, y_i)$, $\mathcal{R}_l^{-i} = \mathcal{R}_l \setminus \{\mathbf{z}_i\}$, and $K^*(\mathbf{z}_i, \mathbf{z}_j) = K_{h_1}(y_i - y_j)K_{h_2}(\mathbf{x}_i - \mathbf{x}_j)$. Clearly, if the value of $K^*(\mathbf{z}_i, \mathbf{z}_j)$ is (approximately) constant for fixed $\mathcal{R}_k$ and $\mathcal{R}_l$, then we can write:

$$CVLL(h_1, h_2) \approx \left(\frac{1}{n} \sum_{k=1}^{m} |\mathcal{R}_k| \log \sum_{l=1}^{m} |\mathcal{R}_l^{-i}| K^*(\mathbf{z}_i, \mathbf{z}_j) \right) - \log(n-1) \qquad (5.65)$$

where $\mathbf{z}_i$ and $\mathbf{z}_j$ are arbitrary elements of $\mathcal{R}_k$ and $\mathcal{R}_l^{-i}$ respectively. While the naive formulation of the CVLL function (as given in Equation 5.60) requires $O(n^2)$ calculations, the complexity of computing the value of Equation 5.65 is instead $O(m^2)$.

Therefore, provided that the feature space is partitioned into a conveniently small number of regions, dual-tree recursion can offer a significant computational advantage over the naive CVLL-based approach.

Dual-tree recursion can be described as consisting of two steps. First, the data points are organized into a kd-tree, where each node in the tree corresponds to a subset $\mathcal{R}$ of $\mathbf{D}$. Second, the kd-tree is traversed simultaneously by means of two loops (with each loop iterating over nodes of the tree). At each iteration of the double traversal, a node pair $(\mathcal{R}_i, \mathcal{R}_j)$ is evaluated in order to check whether the value of a specified approximation of Equation 5.64 lies within sufficiently tight error bounds. If the estimation error is sufficiently small, the estimated value is used for computing the overall objective function, otherwise the approximation is refined by repeating the estimate for subsets of $\mathcal{R}_i$ and $\mathcal{R}_j$. We now present a method for addressing these two tasks respectively.

We first provide an algorithm for organizing the dataset $\mathbf{D}$ into a kd-tree. A kd-tree is nothing but a binary tree $\mathcal{G} = (\mathcal{V}, \mathcal{E})$ such that the root node contains all data points in $\mathbf{D}$, and for each edge $(\mathcal{R}_i, \mathcal{R}_j)$ in $\mathcal{E}$, $\mathcal{R}_j \subseteq \mathcal{R}_i$. The algorithm works recursively, by splitting each node $\mathcal{R}_i$ (starting from the root) into two children $\mathcal{R}_j$ and $\mathcal{R}_k$ such that $\mathcal{R}_i = \mathcal{R}_j \cup \mathcal{R}_k$ and $\mathcal{R}_j \cap \mathcal{R}_k = \emptyset$. The criterion for assigning each point in $\mathcal{R}_i$ to either $\mathcal{R}_j$ or $\mathcal{R}_k$ is the following. If each data point $\mathbf{z}$ is a vector $(z_1, \ldots, z_d)$, we take the dimension along which points in $\mathcal{R}_i$ display the maximum spread. That is to say, we consider the dimension l that maximizes the difference $\max_{\mathbf{z} \in \mathcal{R}_i} z_l - \min_{\mathbf{z} \in \mathcal{R}_i} z_l$. Now, given dimension l, consider the segment $\mathcal{S}$ along l whose endpoints are given by the maximum and minimum values of variable Z_l over $\mathcal{R}_i$. If the midpoint of $\mathcal{S}_l$ is denoted by mid_l, we then assign $\mathcal{R}_j$ all points $\mathbf{z}$ such that $z_l \leq mid_l$, while $\mathcal{R}_k$ is assigned all points in $\mathcal{R}_i \setminus \mathcal{R}_j$. The algorithm stops when all nodes containing at least two points have been assigned a pair of children, so that each leaf in the tree contains exactly one data point (see Algorithm 5.2).

Given the kd-tree, we need to specify a technique both for computing a fast approximation of Equation 5.64 and for estimating the error involved in the approximation. Let $S(\mathbf{z}_i, \mathcal{R}_j)$ be defined as follows:

$$S(\mathbf{z}_i, \mathcal{R}_j) = \sum_{\mathbf{z}_k \in \mathcal{R}_j^{-i}} K^*(\mathbf{z}_i, \mathbf{z}_k) \tag{5.66}$$

Moreover, let $\Delta f(x)$ denote the absolute error in our estimate $\hat{f}(x)$ of a quantity $f(x)$. That is to say, $\Delta f(x) = |\hat{f}(x) - f(x)|$. Our goal is then twofold. First, we need to compute a scalable estimate $\widehat{S}(\mathbf{z}_i, \mathcal{R}_j)$ of the function $S(\mathbf{z}_i, \mathcal{R}_j)$, for any $\mathbf{z}_i$ belonging to a given region $\mathcal{R}_k$. Second, we need to verify whether the value of $\Delta CVLL(h_1, h_2)$ lies within a specified error bound ε.

The key to understanding how the mentioned problems are solved using Monte Carlo sampling is provided by a couple of relevant lemmata. The first lemma specifies a sufficient condition for meeting the error bound imposed on the CVLL function:

Algorithm 5.2 `buildKDTree`: *kd*-tree construction for dual-tree recursion

---

Input: Dataset $\mathbf{D} = \{\mathbf{z}_1, \ldots, \mathbf{z}_n\}$, where $\mathbf{z}_i = (z_{i_1}, \ldots, z_{i_d})$.
Output: *kd*-tree $\mathcal{G} = (\mathcal{V}, \mathcal{E})$.

```
buildKDTree(D):
 1.     V = ∅
 2.     E = ∅
 3.     R₀ = {zᵢ : 1 ≤ i ≤ n}
 4.     ρ = {R₀}
 5.     while(ρ ≠ ∅)
 6.         Rᵢ = an arbitrary element of ρ
 7.         V = V ∪ {Rᵢ}
 8.         if(|Rᵢ| > 1)
 9.             j = 0
10.             maxSpread = 0
11.             for(k = 1 to d)
12.                 spread = maxz∈Rᵢ zk − minz∈Rᵢ zk
13.                 if(spread > maxSpread)
14.                     maxSpread = spread
15.                     j = k
16.             Rk = ∅
17.             mid = minz∈Rᵢ zj + ½(maxz∈Rᵢ zj − minz∈Rᵢ zj)
18.             for(z ∈ Rᵢ)
19.                 if(zj ≤ mid)
20.                     Rk = Rk ∪ {z}
21.             Rl = Rᵢ \ Rk
22.             V = V ∪ {Rk, Rl}
23.             E = E ∪ {(Rᵢ, Rk), (Rᵢ, Rl)}
24.             ρ = ρ ∪ {Rk, Rl}
25.         ρ = ρ \ {Rᵢ}
26.     return (V, E)
```

Lemma 5.3. *Consider some partitioning $\rho = \{\mathcal{R}_1, \ldots, \mathcal{R}_m\}$ of the dataset $\mathbf{D} = \{\mathbf{z}_1, \ldots, \mathbf{z}_n\}$, and let ε denote a specified bound on the error $\Delta CVLL(h_1, h_2)$. Then, if $\frac{\Delta S(\mathbf{z}_i, \mathcal{R}_j)}{S(\mathbf{z}_i, \mathcal{R}_j)} \leq e^{\varepsilon} - 1$ for $1 \leq i \leq n$ and $1 \leq j \leq m$, it follows that $\Delta CVLL(h_1, h_2) \leq \varepsilon$.*

Proof. See e.g. [139].   □

Based on Lemma 5.3, our aim is then to devise a method for estimating $S(\mathbf{z}_i, \mathcal{R}_j)$ such that the condition $\frac{\Delta S(\mathbf{z}_i, \mathcal{R}_j)}{S(\mathbf{z}_i, \mathcal{R}_j)} \leq e^{\varepsilon} - 1$ is satisfied. The Monte Carlo approach to dual-tree recursion proceeds as follows. If we want to estimate the values of $S(\mathbf{z}_i, \mathcal{R}_j)$ for a pair of nodes $(\mathcal{R}_k, \mathcal{R}_j)$, we first construct (via bootstrapping [78]) a number b of samples $\mathcal{S}_1, \ldots, \mathcal{S}_b$ such that, for $1 \leq l \leq b$, $\mathcal{S}_l \subseteq \mathcal{R}_k \times \mathcal{R}_j$ and $\mathcal{S}_l$ contains s pairs of data points $(\mathbf{z}_i, \mathbf{z}_m)$ uniformly sampled from $\mathcal{R}_k \times \mathcal{R}_j$, with $i \neq m$. Second, we estimate the mean μ_{K^*} of $K^*(\mathbf{z}_i, \mathbf{z}_m)$ over the samples in terms of the average value $\hat{\mu}_{K^*}$, defined as follows:

$$\hat{\mu}_{K^*} = \frac{1}{bs} \sum_{l=1}^{b} \sum_{(\mathbf{z}_i, \mathbf{z}_m) \in \mathcal{S}_l} K^*(\mathbf{z}_i, \mathbf{z}_m) \tag{5.67}$$

Third, we compute the standard error $\hat{\sigma}_{K^*}$ of $\hat{\mu}_{K^*}$ over the bootstrap resamplings $\mathcal{S}_1, \ldots, \mathcal{S}_b$, that is:

$$\hat{\sigma}_{K^*} = \sqrt{\frac{1}{b} \sum_{l=1}^{b} \left\{ \left(\frac{1}{s} \sum_{(\mathbf{z}_i, \mathbf{z}_m) \in \mathcal{S}_l} K^*(\mathbf{z}_i, \mathbf{z}_m) \right) - \hat{\mu}_{K^*} \right\}^2} \tag{5.68}$$

Finally, we estimate $S(\mathbf{z}_i, \mathcal{R}_j)$ as:

$$\widehat{S}(\mathbf{z}_i, \mathcal{R}_j) = s\,\hat{\mu}_{K^*} \tag{5.69}$$

Then, we can verify (with a certain confidence degree [77, 76]) whether the value of $\Delta CVLL(h_1, h_2)$ falls below the bound ε by appealing to the following lemma:

Lemma 5.4. *Let C_z denote the probability associated with a normal confidence interval for $\hat{\mu}_{K^*}$ of width $2z$. Then, the condition that $\frac{\Delta S(\mathbf{z}_i, \mathcal{R}_j)}{S(\mathbf{z}_i, \mathcal{R}_j)} \leq e^{\varepsilon} - 1$ is satisfied with probability C_z if $\frac{z\hat{\sigma}_{K^*}}{\hat{\mu}_{K^*}} \leq e^{\varepsilon} - 1$.*

Proof. See e.g. [139]. $\qquad\qquad\qquad\qquad\qquad\qquad\qquad\qquad\qquad\qquad\qquad\qquad$ □

A formal statement of the described CVLL estimation technique is provided by Algorithm 5.3.

Note that the parameters ε, s, b, and z need to be specified by the user. While the choice for ε presumably depends on the specific nature and aims of the application, it is reasonable to wonder whether there may be a generally good choice for the remaining parameters. As a rule of the thumb, the suggestion made in [138, 139] is to set $s = 25$, $b = 10$, and $z = 1.5$, since these values provide a fairly good tradeoff between accuracy and speed. An important remark concerns the sample size s. At first glance, the suggested setting for this parameter may seem to be too small when dealing with very large datasets. However, one has to bear in mind that at higher nodes in the kd-tree (i.e. for larger regions of the feature space) the Monte Carlo estimates are not expected to provide an accurate value of $\hat{\mu}_{K^*}$, but they are only expected to indicate whether the standard error of $\hat{\mu}_{K^*}$ is large enough to motivate additional node splits.

5.6 Structure Learning

Although the original version of Markov Blanket Merging (Section 4.6) is designed for learning discrete hybrid random fields, little modification is sufficient in order to use it for estiamating the structure of continuous hybrid random fields. Sections 5.6.1–5.6.3 describe the way we modify MBM in order to adapt it to continuous domains.

Algorithm 5.3 MCDualTreeCVLL: Monte Carlo dual-tree recursion for CVLL estimation

Input: Dataset $\mathbf{D} = \{\mathbf{z}_1,\ldots,\mathbf{z}_n\}$, where $\mathbf{z}_i = (\mathbf{x}_i, y_i)$; bandwidths h_1 and h_2; error bound ε; sample size s; number b of bootstrap resamplings; number z of standard errors.
Output: Approximate estimate of $CVLL(h_1, h_2)$.

```
MCDualTreeCVLL(D, h₁, h₂, ε, s, b, z):
 1.    for(i = 1 to n)
 2.        Ŝᵢ = 0
 3.    (V, E) = buildKDTree(D)  //See Algorithm 5.2
 4.    R₀ = the root node of 𝒢
 5.    ρ = {(R₀, R₀)}
 6.    while(R ≠ ∅)
 7.        (Rᵢ, Rⱼ) = an arbitrary element of ρ
 8.        if(|Rᵢ|·|Rⱼ| ≤ s)
 9.            for(zₖ ∈ Rᵢ)
10.                Ŝₖ = Ŝₖ + S(zₖ, Rⱼ)  //See Equation 5.66
11.        else if(z σ̂_{K*}/μ̂_{K*} ≤ eᵉ − 1)  //See Equations 5.67–5.68
12.            for(zₖ ∈ Rᵢ)
13.                Ŝₖ = Ŝₖ + Ŝ(zₖ, Rⱼ)  //See Equation 5.69
14.        else
15.            Rᵢ* = {Rₖ : (Rᵢ, Rₖ) ∈ E}
16.            Rⱼ* = {Rₖ : (Rⱼ, Rₖ) ∈ E}
17.            ρ = ρ ∪ (Rᵢ* × Rⱼ*)
18.        ρ = ρ \ {(Rᵢ, Rⱼ)}
19.    return (1/n ∑ⁿᵢ₌₁ log Ŝᵢ) − log(n − 1)
```

5.6.1 Model Initialization

One part of the algorithm that needs to be modified in a suitable way is the model initialization technique. In discrete HRFs, MBM produces an initial assignment by choosing an initial size k of the set of relatives, and then by selecting as relatives of each X_i the k variables that display the highest statistical correlation with respect to X_i, where the strength of the correlation is measured by the value of the χ^2 statistic. Since the χ^2 statistic naturally applies to discrete variables only, what we need is a way of measuring correlation for pairs of continuous variables in a direct way (i.e. without having to discretize the variables before applying the test).

Our choice is to measure the statistical correlation for any pair of continuous variables by the value of the correlation ratio [164] for that pair. Consider two random variables X_i and X_j that have been observed n times within a dataset $\mathbf{D}$. Moreover, define $\hat{\mu}_i$, $\hat{\mu}_j$, and $\hat{\mu}$ in the following way:

$$\hat{\mu}_i = \frac{1}{n} \sum_{k=1}^{n} x_{i_k} \tag{5.70}$$

$$\hat{\mu}_j = \frac{1}{n} \sum_{k=1}^{n} x_{j_k} \tag{5.71}$$

$$\hat{\mu} = \frac{1}{2}(\hat{\mu}_i + \hat{\mu}_j) \tag{5.72}$$

where x_{i_k} and x_{j_k} denote the values of the k-th observation of x_i and x_j in $\mathbf{D}$, respectively. Then, the correlation ratio statistic η for the pair (X_i, X_j) can be computed as follows:

$$\eta(X_i, X_j) = \sqrt{\frac{n(\hat{\mu}_i - \hat{\mu})^2 + n(\hat{\mu}_j - \hat{\mu})^2}{\sum_{k=1}^{n} \left\{ (x_{i_k} - \hat{\mu})^2 + (x_{j_k} - \hat{\mu})^2 \right\}}} \tag{5.73}$$

The correlation ratio is such that $0 \leq \eta \leq 1$, where lower values correspond to stronger degrees of correlation, while higher values mean weaker correlation. One advantage of using the correlation ratio statistic is that it is a fairly general dependence test, capable of detecting not only linear dependencies but also non-linear ones. On the other hand, most traditional dependence tests (such as Pearson's correlation coefficient [165]) only capture linear dependencies. Therefore, the correlation ratio is a suitable choice for the initialization of Markov Blanket Merging, since it allows to estimate densities without making assumptions on the nature of the modeled dependencies.

5.6.2   Learning the Local Structures

In the Bayesian networks composing continuous hybrid random fields, the conditional density of each node given its parents is modeled by using either parametric, semiparametric, or nonparametric estimators (as described in Sections 5.3–5.5). For root nodes, conditional density estimation clearly reduces to standard (unconditional) pdf estimation. An important issue with Markov Blanket Merging that needs to be addressed in a different way when dealing with continuous domains is the scoring function used to evaluate the structure of the local Bayes nets. While the original version of MBM uses a heuristic function based on the minimum description length principle (see Section 4.6), the evaluation function we suggest for continuous graphical models is the model CVLL with respect to the training dataset $\mathbf{D}$ [136, 137, 97]. For a BN with graph $\mathcal{G}$ and nodes $X_1, \ldots, X_d$, if $\mathbf{D} = \{\mathbf{x}_1, \ldots, \mathbf{x}_n\}$ and each $\mathbf{x}_j$ is a vector $(x_1, \ldots, x_d)$, the structure $\mathcal{G}$ is scored as follows:

$$CVLL(\mathcal{G}) = \sum_{j=1}^{n} \sum_{i=1}^{d} \log \hat{p}^{-j} \left(x_{i_j} \mid pa_j(X_i) \right) \tag{5.74}$$

where $pa_j(X_i)$ is the state of the parents of X_i in $\mathbf{x}_j$, and the notation $\hat{p}^{-j}$ means that only data points in $\mathbf{D} \setminus \{\mathbf{x}_j\}$ are used as training sample for estimating the pdf of point $\mathbf{x}_j$. Clearly, a CVLL-based strategy is much less prone to overfitting than a straight maximum-likelihood approach. The CVLL function is maximized (up to a local optimum) by heuristic search in the space of d-dimensional BN structures.

To this aim, we use (as for discrete hybrid random fields) the greedy hill-climbing algorithm described in Section 2.4.2.1.

5.6.3   Learning the Global Structure

The last correction we need to introduce in the Markov Blanket Merging algorithm concerns the evaluation function used for scoring the global structure of the hybrid random field. Rather than maximizing straightly the pseudo-log-likelihood of the model given the dataset, we suggest to optimize instead a cross-validated version of that function, consistently with the choice we made also for local structure learning. For a dataset $\mathbf{D}$ containing n d-dimensional patterns and a HRF with graphs $\mathcal{G}_1, \ldots, \mathcal{G}_d$, the cross-validated pseudo-log-likelihood measure, denoted by $CVLL^*(\mathcal{G}_1, \ldots, \mathcal{G}_d)$, is defined by the following equation:

$$CVLL^*(\mathcal{G}_1, \ldots, \mathcal{G}_d) = \sum_{j=1}^{n} \sum_{i=1}^{d} \log \hat{p}^{-j}(x_{i_j} \mid mb_{i_j}(X_i)) \tag{5.75}$$

where $mb_{i_j}(X_i)$ is the state of the MB of X_i in pattern $\mathbf{x}_j$.

5.7   Final Remarks

A long-lasting debate has been going on in the scientific community on whether continuous-valued variables do offer advantages over discrete attributes or not. Although the ultimate solution to the argumentation is way beyond the scope of the book, we brought in several arguments in order to stress the importance of developing graphical models for continuous variables. Most of the formal pros stem from the analysis of optimal convergence of the Parzen window technique for the estimation of a density function. Furthermore, the definition of different versions of hybrid random fields for continuous features allows for a fair, straightforward experimental comparison with respect to traditional extensions of Bayesian networks and Markov random fields to real-valued setups. In fact, a number of datasets and real-world application domains are collected and represented in the form of continuous attributes in quite a natural manner.

Generalization of the graphical paradigm to real-valued random vectors did not affect the formal definition and the theoretical properties of the hybrid random field. Adaptation of the discrete model to the new scenario took place at two levels, namely the adoption of a specific model for the posterior probabilities of continuous variables given their Markov blankets, and consequent adoption of suitable estimation algorithms for the involved distributions. We investigated three different approaches to the problem of estimating the conditional probability density functions associated with the nodes of the graphical model, i.e. parametric, semiparametric, and nonparametric.

Parametric estimation relies on the notion of possessing some prior knowledge on the form of the underlying density function, such that the aim and scope of learning from data is limited to the estimation of a set of parameters that characterize uniquely the pdf itself. The treatment we gave benefited from the classic discussion on Normal and Gaussian mixture models, where the maximum-likelihood technique emerges as a sound paradigm for finding out an exact (i.e., in closed-form) or iterative solution to the estimation problem. Parametric estimation has its strongest point in its relative simplicity, which may entail robustness. Once the parameters have been estimated, the resulting model is also computationally fast. Furthermore, the GMM is also corroborated by its theoretical 'universality', that is its capability of modeling any continuous pdf to any degree of precision (provided that the correct parameters and number of components are used). Unfortunately, this property is of little practical help, since (*i*) the correct GMM for approximating the pdf underlying a given dataset is not known, (*ii*) it may require a huge number of components, and (*iii*) iterative ML estimation is unlikely to return the optimal parameters (even in the lucky event that it is applied to the ideal mixture). Eventually, the main cutoff of parametric statistics lies in its requirement of a priori knowledge of the form of the pdf, which turns out to be too strong an assumption.

A semiparametric approach was outlined, too, relying on the change of variables theorem and the intriguing definition of nonparanormal distribution. Efforts were concentrated on the computation of suitable mappings of the original feature space, mappings that entail the interpretation (and the handling) of the data as a population drawn from an adequate nonparanormal distribution. While parametric models fix the form of the pdf, semiparametric techniques aim at transforming any probability distribution into a reference density function whose form is actually known. This may, or may not, fit specific scenarios well, depending on the specific data. At any rate, the approach shifts the emphasis (including the fitness of the resulting model, and a critical mass of the computational burden) from explicit density estimation to designing the transformation functions that realize the change of variables.

Finally, fully nonparametric models were applied. They relied on the extension of the Parzen window technique to the conditional density estimation setup. The resulting model, known as the Nadaraya-Watson estimator, involves two separate kernels having (possibly) different bandwidths. In order to ease the (non-trivial) task of selecting proper bandwidth values, an effective model selection algorithm based on the cross-validated log-likelihood criterion was presented. Due to the typical complexity issues that may arise in the computation of likelihood-based measures on large-size datasets, a workaround using dual-tree recursion was described. The selling-points of nonparametric estimation techniques lie in: (*i*) their optimal convergence to the real pdf over large datasets; (*ii*) the nonoccurrence of prior assumptions on the form of the unknown pdfs; (*iii*) the contained amount of parameters that need to undergo model selection heuristics (basically limited to the kernel bandwidths); (*iv*) the absence of a real training stage, since the model is simply memory-based and the calculations occur only at test time. The last two points hide also the germ of the severe drawbacks of nonparametric techniques: bandwidth selection is difficult and time consuming, and the very nature of memory-based paradigms (that are required to store the whole training sample in memory, and to compute a linear combination of as many kernel calculations as the number of training points) may prevent the resulting machine from scaling up to large-size datasets in a satisfying way.

Once a model of the conditional density functions associated with the vertexes of the hybrid random field has been chosen, structure learning can be applied. We saw how the problem of learning the conditional independencies to be modeled by HRFs for continuous variables is best tackled by subdividing it into three phases, namely initialization, 'local' structure learning, and 'global' structure learning. The next chapter puts real HRFs at work, evaluating their behavior in several applications involving either discrete or continuous attributes.

Chapter 6
Applications

"[A] dire il vero ho poca inchinazione al filosofare altamente;
ed ho anche stimato che non v'abbisogni una grande sublimità
d'intelletto de' discorsi che hanno per meta l'intenzione di
scoprire la pura e semplice verità sotto gl'insegnamenti del
senso; e se questi m'ha ingannato, a chi doveva io ricorrere?"

Agostino Scilla, 1670 [272]

6.1 Introduction

As we have said, hybrid random fields are not meant just as a general graphical
model with nice theoretical properties, featuring algorithms for inference and learning over discrete and continuous variables. Above all, they are expected to reveal
useful. This means that our ultimate goal is to exploit the flexibility of HRFs in
modeling independence structures, as well as the scalability of algorithms for learning HRFs, in order to tackle real-world problems. Improvements over the traditional
approaches, both in terms of prediction accuracy and computational efficiency, are
sought. This chapter presents and analyzes a variety of applications, confirming the
expectations to a significant extent. Experimental results yielded by HRFs turn out to
be at least as accurate as those obtained via long established paradigms. Moreover,
the computational burden proves to be invariably reduced, even to dramatic extents,
demonstrating that hybrid random fields are a viable tool for designing serious applications whenever the size of the dataset scales up to severe dimensionality. All
models and algorithms considered in the applications are publicly available through
the implementation provided by the JProGraM software library, which is released at
http://jprogram.sourceforge.net/ under an open-source license. Also,
the datasets used in the demonstrations that are not otherwise available on the web
are hosted at http://www.dii.unisi.it/~freno/datasets.html.

We present a feature selection (or, dimensionality reduction) technique relying
on graphical models first (Section 6.2). The idea is that the feature space used for
representing real-world phenomena may be redundant, or it may have too high a
dimensionality. Under these circumstances, a selection of the most relevant features
may help tackling the 'curse of dimensionality' issue [73]. Two requirements need
to be satisfied by the selection procedure, namely that a significant reduction of the
dimensionality is obtained, and that the performance of the resulting (simplified)
machine is not degraded. Section 6.2.1 shows how probabilistic graphical modeling may be suitably applied to the task, in the form of a technique called 'Markov

A. Freno and E. Trentin: Hybrid Random Fields, ISRL 15, pp. 121–150.
springerlink.com   © Springer-Verlag Berlin Heidelberg 2011

blanket filter'. Alternative, related work is reviewed in Section 6.2.2. The Markov blanket filter technique is then used, in conjunction with a few traditional classifiers, over four different classification tasks (including two bioinformatics datasets) in Section 6.2.3. Results show that the proposed approach yields, on average, a recognition accuracy which is comparable to (or improved over) established methods, still marking an even more noticeable dimensionality reduction.

Section 6.3 discusses the application of HRFs to pattern recognition and link prediction problems involving discrete variables. Section 6.3.1 preliminarily describes the strategies used for setting up the structure (and learning the parameters) of Markov random fields and dependency networks in the absence of prior knowledge concerning the application domain. Complexity issues are investigated first (Section 6.3.2), evaluating the computational burden of structure learning as a function of the growing size of synthetic datasets. It turns out that learning HRFs scales up (even dramatically) better than learning Bayesian and Markov networks. Section 6.3.3 reports on several pattern recognition experiments (including the popular Lung-Cancer dataset) where the HRF is used as a classifier, and the focus of the analysis is on the prediction accuracy. Again, HRFs compare favorably with respect to the other graphical models, and with respect to the naive Bayes classifier as well. A domain having the utmost relevance to World Wide Web-related applications lies in the so-called link prediction setup (Section 6.3.4). We point out how HRFs can be used for link prediction according to a specific ranking strategy (Section 6.3.4.1), then we apply the model to the task of predicting references in scientific papers from the well-known CiteSeer and Cora datasets (Section 6.3.4.2). The results, evaluated in terms of the 'mean reciprocal rank' and the 'success rate at N' criteria, highlight the benefits marked by HRFs. Another popular link prediction task, useful in the development of automatic recommender systems for the World Wide Web, concerns the prediction of users preferences for certain families of products or services (Section 6.3.4.3). Preferences for movies from the MovieLens dataset are considered here, and the models are evaluated in terms of two variants of the 'degree of agreement' (DOA) metric, namely the macro-averaged DOA and the micro-averaged DOA. Again, experimental results prove that HRFs compare favorably, on average, with the traditional approaches.

Section 6.4 applies HRFs to pattern recognition tasks involving continuous-valued variables. The algorithms introduced in the previous chapter for estimating conditional pdfs over real-valued feature vectors are evaluated and compared on complex datasets generated synthetically. Section 6.4.1 hands out the algorithms used for creating the datasets, a process which involves three separate steps, namely: (*i*) generation of a random DAG, (*ii*) generation of a random, multivariate pdf (featuring polynomial dependencies among the variables, plus noise distributed according to various beta densities) from a given DAG, and (*iii*) creation of pattern-classification datasets from a collection of class-conditional generative graphical models (relying on Monte Carlo sampling, as applied to the output of the previous two algorithms). Experiments are reported and analyzed, in terms of both recognition accuracy and average training time, in Section 6.4.2. The results allow us to extend to the continuous-valued scenario the positive conclusions we drew for HRFs in the discrete case. Final remarks are given in Section 6.5.

6.2 Selecting Features by Learning Markov Blankets

Typical supervised classification tasks involve representing the patterns in a dataset as vectors of features. Algorithms that learn to classify patterns are then applied to such vectors. In complex domains, the number of features can be very large [73], and the vector size has an important impact on the performance of the algorithms, for at least two reasons. First, high-dimensional vectors increase both the amount of space and the time required for learning. Second, when the vectors also include irrelevant variables, the presence of such variables may degrade the accuracy of the learned classification model. Therefore, reducing the dimensionality of vectors may improve the performance of a learning algorithm in two respects. First, it can decrease the complexity of the learning process. Second, it can lead the learning algorithm to induce more reliable models of the domain at hand. One way of addressing dimensionality reduction in vectors of features is through *feature selection* [122].

Feature selection techniques can be classified into two broad families [153, 170]: *filters* and *wrappers*. The distinction between filters and wrappers depends on the relationship they bear to the classification algorithm. Filters select features based on criteria which are independent of the particular learning algorithm to be applied to the data. Hence, the bias guiding feature selection in filters is not related to the learning bias of the classifier. In wrappers, a learning algorithm is first chosen, and then the classification accuracy on different subsets of features is used as a heuristic for selecting the best subset.

Wrapper methods have both a practical shortcoming and a theoretical limitation [176]. On the one hand, they typically incur a high computational cost, since a classifier is required to be trained several times on the same task, with different subsets of features. On the other hand, they do not permit to investigate the problem of dimensionality reduction as such. Since features are selected only insofar as classification accuracy is increased, wrappers obscure the nature of the interaction between dimensionality reduction and learning accuracy. While the first limitation of wrapper techniques can be overcome by devising more efficient algorithms [122], the second limitation is more general, since it depends on the very nature of these techniques.

6.2.1 A Feature Selection Technique

We now describe a filter method based on Bayesian networks, referred to as *Markov Blanket Filter* (MBF), which was first proposed in [93]. The basic idea consists in modeling the features, together with the label attached to the vector, as a set of (discrete) random variables, and then learning the joint probability distribution of the variables in the form of a Bayesian network. Once a Bayesian network has been learned, the Markov blanket of the label within the network specifies a subset of the features that makes the label independent of all other features. Therefore, within the framework of Bayesian networks (and, in general, within the framework of probabilistic graphical models), feature selection can be reduced to the task of extracting a Markov blanket for the class from the set containing all the original variables.

Consider the traditional task of supervised learning, where the training data consists of feature vectors $\mathbf{X} = (X_1, \ldots, X_n)$ such that the label C is known for each vector. Clearly, each feature X can be regarded as a random variable ranging over a fixed domain of possible values. The same is true for the label C, which can be treated as a random variable ranging over the set of possible labels $c_1, \ldots, c_m$. If we take as training data the set of vectors $\mathbf{X}^* = (X_1, \ldots, X_n, C)$, we may reduce the problem of feature selection to the following one: what is the Markov blanket of C in $\{X_1, \ldots, X_n\}$? Since a Bayesian network implicitly specifies a Markov blanket for every variable contained in the DAG, the latter problem can be solved by learning a Bayesian network for the probability distribution underlying the labeled data. In practice, any algorithm capable of learning the structure of Bayesian networks can be used as a feature selection algorithm. The MBF algorithm learns a Bayesian network for the dataset by running the hill-climbing algorithm described in Section 2.4.2.1 (i.e. Algorithm 2.1), and then it picks out the features belonging to the Markov blanket of the class variable.

6.2.2   Related Work

The idea of applying the notion of Markov blanket to feature selection was first proposed in [176]. The proposed algorithm requires the user to specify in advance some important parameters, in particular the number of features that have to be eliminated. But since in typical applications the most sensible value for that number is unknown, this limitation either leads to suboptimal results (when the user is not able to choose a sensible number), or makes it necessary to run the algorithm several times (setting that number each time to a different value, until good results are obtained). On the other hand, MBF does not require the user to set any parameters at all when it is applied to new data.

In [143], feature selection is also dealt with by means of Bayesian networks. However, the learning algorithm employed in this case, called 'K2χ^2', has two serious drawbacks. K2χ^2 is based on the K2 algorithm (described in Section 2.4.2.2), which requires the user to specify two parameters: first, an ancestral ordering over the network nodes; second, an upper bound on the number of parents allowed for each node. Concerning the first parameter, K2χ^2 orders the variables according to the strength of the dependence relationship between each one of them and the class variable, measuring this dependence through the χ^2 test. Thus, variables with higher

χ^2 scores precede variables with lower scores. While [143] show that the resulting ordering works better than a random ordering, no connection is established between the χ^2 test and the semantics of the node ordering in Bayesian networks. This leaves room for some doubts as to how reliable the resulting ordering will be. Concerning the second parameter, no method is provided by the authors for specifying it. This means that the user will have to provide the algorithm with the right parameter setting each time it is applied to new data, and there seems to be no general criterion for going about this task. On the other hand, the structure learning algorithm used by MBF requires neither to specify a variable ordering, nor to fix an upper bound on the number of parents allowed for each node, and this makes MBF both more general and more appealing for application to domains where no prior knowledge can be used to constrain the learning process.

6.2.3 Results

MBF was tested on four different datasets, concerning one synthetic and three real-world domains, which we describe next:

1. The LED24 dataset [230] is a synthetic one, containing 3200 patterns. Each pattern is characterized by 24 boolean features, and it belongs to one of 10 possible classes. Only 7 features are relevant for predicting the class, and the class entirely determines the value of each relevant feature (except for a 0.1 probability that the value of the feature will be chosen randomly, introducing a certain noise). The values of the other 17 features are random.
2. The Mushroom dataset [230] contains 8124 patterns. Each pattern describes a mushroom through 22 features. All features are nominally valued, and range over domains containing up to 12 possible values. The class variable has two possible values: either 'edible' or 'poisonous'.
3. The Splice-junction dataset [230] consists of 3190 patterns. The patterns are characterized by 60 four-valued attributes, where each attribute stands for a position on a DNA sequence. The task is to recognize whether the pattern is an exon/intron boundary, an intron/exon boundary, or neither.
4. The HS3D dataset has been extracted from the Homo Sapiens Splice Sites database [245]. The dataset was prepared by merging the Exon-Intron true splice sites data from that database with the first 1500 items of the Exon-Intron false splice sites data. In this way, we obtain a dataset containing 4296 patterns, belonging to two possible classes: either true or false exon/intron splice site. Each pattern is a vector of 140 features, describing a DNA sequence; as for the preceding Splice-junction dataset, each feature is a four-valued variable, standing for the nucleotide occupying the specific position in the sequence.

The strategy used to test MBF on each dataset is the following: (1) train some classifiers on the dataset, and record their classification accuracy; (2) run MBF on the data; (3) train the classifiers used in step 1 on the dataset obtained by eliminating from the original data the features discarded in step 2, and compare their classification accuracy to the accuracy recorded in step 1. Classification accuracy is measured using ten-fold cross-validation. The employed classification algorithms are the following (as supplied by the WEKA machine learning workbench [317]): naive Bayes; J4.8, which is WEKA's implementation of the C4.5 algorithm [249]; IB1 [3]. These specific algorithms are chosen so as to make it possible to evaluate the impact of MBF across a variety of classification techniques (i.e. Bayesian classifiers, tree classifiers, and nearest-neighbor classifiers, respectively). The results of MBF are compared to the results of the CFS algorithm [123], which is a well-known and efficient feature selection method (also supplied by WEKA) and hence provides an interesting baseline for evaluating MBF.

The results are summarized in Tables 6.1–6.2. Table 6.1 compares, for each dataset, the number of features selected by MBF to the number of features selected by CFS. Table 6.2 shows the classification accuracy of the three classifiers for each dataset, comparing their accuracy on the original datasets to the accuracy on the datasets filtered by MBF and by CFS respectively. Concerning the number of features selected by the two algorithms, MBF seems to perform better: summing over all four cases, CFS ends up selecting 5 features more than MBF. A strong difference emerges from the LED24 dataset, which is the only synthetical domain. Since we know in advance which features are relevant, this dataset provides a preliminary benchmark for MBF. The result is particularly encouraging, since the selected features for MBF are exactly the relevant ones, whereas CFS selects 8 irrelevant features in addition to the relevant ones. MBF almost always improves the classification accuracy, regardless of the chosen classifier: in only 1 out of 12 cases (i.e. Naive Bayes on the LED24 data) the accuracy decreases after filtering, and the decrease is quite small (0.06%). The accuracy after using CFS decreases instead in 3 cases (ranging from a 0.99% to a 6.11% decrease). In 9 cases, the classification accuracy achieved after filtering the data with MBF is higher than the one achieved by CFS; in 1 case the resulting accuracy is the same (J4.8 on the Splice-junction dataset); in the remaining 2 cases CFS performs better than MBF. The maximum increase in accuracy produced by filtering the data with MBF is 15.22%, while for CFS this value amounts to 12.19% (both values are recorded on LED24 with IB1).

Table 6.1 An overview of the datasets, with the last two columns reporting the number of features selected by MBF and CFS respectively

Dataset	Features	Labels	Patterns	MBF Features	CFS Features
LED24	24	10	3200	7	15
Mushroom	22	2	8124	10	4
Splice	60	3	3190	21	24
HS3D	140	2	4296	21	21

Table 6.2 Average recognition accuracy of the naive Bayes, J4.8, and IB1 classifiers on the original data and on the data filtered by MBF and CFS respectively

	Average Recognition Accuracy (10-fold cross-validation)								
	Naive Bayes			J4.8			IB1		
Dataset	/	MBF	CFS	/	MBF	CFS	/	MBF	CFS
LED24	75.43%	75.37%	75.87%	73.46%	75.56%	74.46%	49.31%	64.53%	61.50%
Mushroom	95.82%	98.05%	98.52%	100%	100%	99.01%	100%	100%	93.89%
Splice	95.29%	96.20%	96.08%	94.38%	94.48%	94.48%	75.92%	81.94%	80.37%
HS3D	83.44%	91.55%	90.17%	91.38%	91.55%	91.03%	76.65%	81.40%	80.60%

6.3 Application to Discrete Domains

This section offers both an illustration of the way that HRFs (and graphical models in general) can be applied to pattern recognition and link prediction tasks, and an experimental investigation of the models behavior in terms of prediction accuracy and computational burden of structure learning. Before presenting the applications, Section 6.3.1 explains how Markov random fields and dependency networks are set up for the considered tasks. We then proceed to an empirical evaluation of the scalability properties of MBM and its capability to accurately learn pseudo-likelihood distributions. Section 6.3.2 is aimed at measuring the computational burden of learning HRFs from data, using a number of synthetic benchmarks. In Section 6.3.3 hybrid random fields are applied to pattern classification, comparing their accuracy to Bayesian networks, Markov random fields, dependency networks, and to the naive Bayes classifier. Section 6.3.4 describes instead the application of graphical models to a number of link prediction tasks, drawn from the widely used CiteSeer, Cora, and MovieLens databases. These datasets display both a rich relational structure and a growing domain dimensionality (in terms of the number of random variables involved in the data). Therefore, the link prediction applications offer an interesting way to evaluate both the learning capabilities of hybrid random fields when dealing with structured domains, and the scalability properties of MBM as the size of the dataset increases.

6.3.1 Setting Up the Markov Random Field and Dependency Network Models

In Markov networks, joint distributions are estimated using the pseudo-likelihood approximation (for the sake of computational efficiency). The model weights are learned by means of a maximum (pseudo-)likelihood strategy. In particular, the model pseudo-likelihood is optimized using the L-BFGS algorithm [196], already mentioned in Section 3.3.1. In order to construct the graph in Markov random fields, we adopt the following strategy. First, for each variable X_i in the domain, we run a

χ^2 test between X_i and each other variable X_j, in order to measure the strength of the correlation existing between X_i and X_j. Then, for each X_i, we select the k variables that achieve the highest scores on the χ^2 test, and we add these k variables as relatives to X_i. In each application, a suitable value for k is determined by preliminary cross-validation. Clearly, the way we construct the graph in Markov networks is very similar to the model initialization step in hybrid random fields.

Concerning Dependency networks, a set of k neighbors is assigned to each node based on the results of the χ^2 test, where k is tuned by cross-validation. The local distributions are then learned using the typical techniques suggested in the relevant literature [128, 100], based on probabilistic decision trees. In all the experiments described below, the parameter k will refer to the number of neighbors (or relatives) initially assigned to each node based on the results of the χ^2 test, both in the case of Markov random fields, dependency networks, and hybrid random fields. On the other hand, the parameter k^* will denote the upper bound on the size of the set of relatives considered when learning hybrid random fields by Markov Blanket Merging.

6.3.2 Computational Burden of Structure Learning

In order to compare the computational cost of learning hybrid random fields to the cost of learning Bayesian networks, Markov random fields, and dependency networks, we measure the time needed to learn the respective models from a number of datasets of growing dimensionality. Each dataset used in the tests contains 1000 n-dimensional patterns, drawn from a single distribution (that is from only one class). All features are binary. The datasets are generated using a rule-based random data generator which is available in the WEKA software package [317].

Given each dataset, we measure the time needed for learning (1) Bayesian networks using Algorithm 2.1, (2) Bayesian networks using the K2 algorithm, (3) dependency networks, (4) hybrid random fields, and (5) Markov random fields using the respective algorithms we described previously. The reason why (for BNs) we compare MBM not only to Algorithm 2.1, but also to K2, is that the computational cost of Algorithm 2.1 prevents us from applying it successfully to the link-prediction experiments, where the domain dimensionality forces us to train the BN models using K2, which is less expensive to run. Clearly, while we are going to learn the structure (and not only the parameters) of BNs and HRFs, for DNs and MRFs the main effort is devoted to parameter estimation, since for the latter models structure learning is limited to the initialization of the neighborhoods. This means that the task is more demanding in the case of Bayesian networks and hybrid random fields. For Markov networks the value of k is set to 6, for DNs it is set to 8, while for HRFs the values of k and k^* are set to 8 and 10 respectively. In the K2 algorithm, likelihood is used as evaluation function and the maximum number of parents allowed for each node is set to 3. We choose these specific parameter values because they are the largest ones we ever considered in our applications, while tuning the parameters in preliminary cross-validation runs of the experiments. Time is measured on a PC equipped with a 2.34 GHz processor. The results are illustrated in Figures 6.1–6.4, where learning time (in seconds) for the different models is plotted against the increasing dimensionality of the data, i.e. for a growing number of variables in the models. The reason why, for these experiments, we do not extend the measurement to datasets involving more than 75 variables is that the chosen interval is sufficient in order to display the differences between the compared curves. That is to say, we do not aim at measuring learning time for large domains as such, but we aim instead at measuring the growth of learning time with respect to a growing number of variables. Moreover, consider that the section describing the link-prediction experiments also reports learning time for all the considered models, providing an indication of how expensive it is to learn the models from domains involving several hundreds of variables.

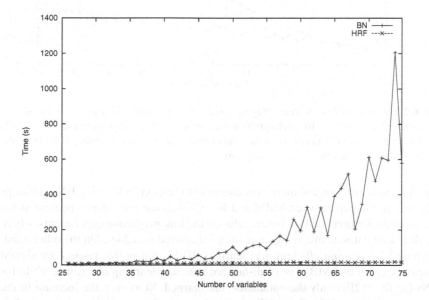

Fig. 6.1 Time required for learning Bayesian networks (using Algorithm 2.1) and hybrid random fields ($k = 8, k^* = 10$) as the problem size increases. For each n such that $26 \leq n \leq 75$, the time (in seconds) is measured with respect to a training set containing 1000 patterns, where each pattern is a vector of n binary variables.

The comparison displays a clear advantage of hybrid random fields over Bayesian networks and Markov random fields. In particular, the improvement of HRFs over BNs (trained with Algorithm 2.1) is dramatic, while the difference between HRFs and MRFs is less consistent. Concerning the K2 algorithm for BNs, training time grows much more quickly than the corresponding time for HRFs. Although K2 is relatively fast for low-dimensional datasets, the time measurements reported in the figures show that as the number of variables increases, learning HRFs (or even

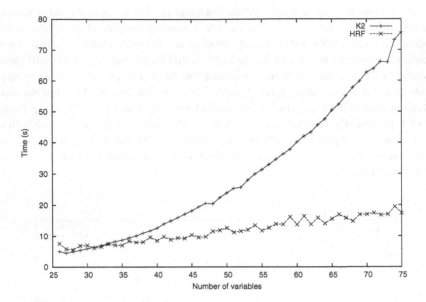

Fig. 6.2 Time required for learning Bayesian networks (using the K2 algorithm) and hybrid random fields ($k = 8, k^* = 10$) as the problem size increases. For each n such that $26 \le n \le 75$, the time (in seconds) is measured with respect to a training set containing 1000 patterns, where each pattern is a vector of n binary variables.

MRFs) becomes more and more convenient with respect to learning BNs. The significance of the gap between MBM and K2 will become even more apparent as we will refer (in Section 6.3.4) to time results for the link-prediction experiments, where we deal with datasets involving more than a thousand variables. On the other hand, learning DNs is more efficient than learning HRFs, but in this respect we should note once again that MBM is a full-fledged structure learning algorithm, while for DNs (as for MRFs) only the parameters are learned. Moreover, the increase in the amount of time needed to learn HRFs with respect to learning DNs is quite small if compared to the amount of time needed to learn BNs or MRFs.

Concerning instead the possible growth of the number of iterations required by MBM for convergence, Figure 6.5 also plots the values of that number against the increasing dimensionality of the data.

6.3.3 Pattern Classification

The experiments described in this section, originally reported in [96], are aimed at evaluating the accuracy of hybrid random fields in pattern recognition. In particular, HRFs are compared not only to Bayesian networks, Markov random fields, and dependency networks, but also to the naive Bayes classifier. Since NB is among

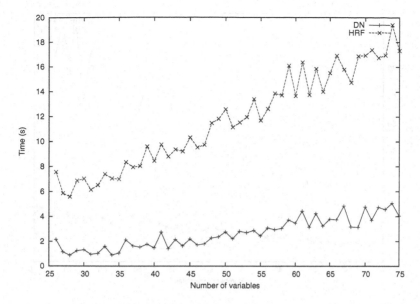

Fig. 6.3 Time required for learning dependency networks ($k = 8$) and hybrid random fields ($k = 8$, $k^* = 10$) as the problem size increases. For each n such that $26 \leq n \leq 75$, the time (in seconds) is measured with respect to a training set containing 1000 patterns, where each pattern is a vector of n binary variables.

the most widely used probabilistic methods in pattern recognition, its classification accuracy in the considered tasks will be a useful benchmark, due to the following fact. Given that the naive Bayes model can be regarded as a suboptimal Bayesian network (i.e. as a BN modeling the conditional independencies among features in a simplified way), BNs should perform better than NB in classification tasks where the features are correlated in important ways. Now, HRFs are aimed at modeling the same kind of complex distributions that are accurately modeled by BNs, rather than the systems of features where NB is capable of achieving a relatively high classification accuracy. Therefore, the experimental question we are going to explore through the pattern recognition tasks is how accurate HRFs are in pattern classification whenever the features are embedded in sufficiently rich networks of conditional independencies.

We now describe the way that BNs, MRFs, DNs, and HRFs are applied to pattern recognition tasks. Suppose $c_1, \ldots, c_n$ are the classes in the problem at hand. We partition the training data $\mathbf{D}$ into n subsets $\mathbf{D}_1, \ldots, \mathbf{D}_n$, such that all patterns in $\mathbf{D}_i$ belong to class c_i. For each probabilistic graphical model, we then learn n class-specific versions, training each version on the respective subset of the training data. Given the model versions $h_1, \ldots, h_n$, in order to classify patterns in the test set we proceed as follows. For each class c_i, we first compute the posterior probability $P(c_i|x)$ that a pattern $\mathbf{x}$ belongs to class c_i:

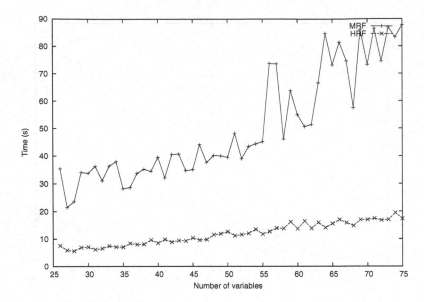

Fig. 6.4 Time required for learning Markov random fields ($k = 6$) and hybrid random fields ($k = 8$, $k^* = 10$) as the problem size increases. For each n such that $26 \leq n \leq 75$, the time (in seconds) is measured with respect to a training set containing 1000 patterns, where each pattern is a vector of n binary variables.

$$P(c_i \mid \mathbf{x}) = \frac{P(\mathbf{x} \mid c_i)P(c_i)}{P(\mathbf{x})} \qquad (6.1)$$

where $P(\mathbf{x} \mid c_i)$ is computed as the likelihood (or the pseudo-likelihood) of model h_i given $\mathbf{x}$ (i.e. as the probability assigned to $\mathbf{x}$ by model h_i), and $P(c_i)$ is the prior probability of class c_i, estimated as $\frac{|\mathbf{D}_i|}{|\mathbf{D}|}$. Given the posterior probability of each class, we attach to $\mathbf{x}$ the label with the highest posterior probability, based on a maximum a posteriori strategy. Clearly, since $P(\mathbf{x})$ is constant for a fixed $\mathbf{x}$, for the purposes of classification it is sufficient to compute $P(\mathbf{x} \mid c_i)P(c_i)$, ignoring the normalization factor $\frac{1}{P(\mathbf{x})}$.

The dataset for the first task, which we call 'Synthesis', is a synthetic one, created using WEKA's rule-based random data generator. The dataset contains 3,000 patterns. Each pattern is characterized by 20 boolean features, and there are two possible classes. The data was generated setting the minimum rule size to 4 (rather than 1, which is the default value), and the maximum rule size to 5, where the size of a rule is the number of variables appearing in the rule formulation. Increasing the minimum rule size makes the task increasingly difficult, because it introduces more complex kinds of dependencies among the variables. This kind of complexity is exactly what we require from the data in order to address our experimental question. Results for the Synthesis dataset are reported in Table 6.3.

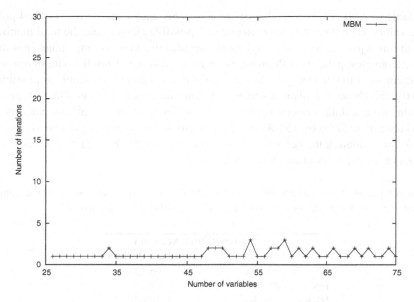

Fig. 6.5 Number of iterations required for Markov Blanket Merging to converge as the problem size increases. For each n such that $26 \leq n \leq 75$, that number is measured with respect to a training set containing 1000 patterns, where each pattern is a vector of n binary variables.

Table 6.3 Recognition accuracy values measured (by 3-fold cross-validation) on the Synthesis dataset for BN, DN ($k = 3$), HRF ($k = k^* = 8$), MRF ($k = 3$), and NB

	Recognition Accuracy	
	Average	Standard Deviation
BN	73.20%	1.55
DN	63.83%	0.74
HRF	73.96%	1.20
MRF	67.03%	0.40
NB	65.36%	0.95

The highest accuracy on the Synthesis data is achieved by HRFs. While the accuracy of BNs is not far from the accuracy of HRFs, the accuracy displayed by MRFs is 6.93 points lower than the value obtained with HRFs. On the other hand, DNs are less accurate than all other models. The fact that NB performs (significantly) worse than BNs confirms the significance of the dataset for testing the reliability of HRFs with respect to domains displaying relatively complex networks of correlations among features.

The second experiment concerns the Lung-Cancer dataset, drawn from the UCI machine learning repository [230]. The dataset is a real-world one, and it was used first by [140] as a particularly illustrative example of an ill-posed pattern recognition setting, where the domain dimensionality overwhelms the number of examples available for training. The number of features is 56, each feature ranges over 4 possible values, each pattern belongs to one of 3 possible classes, and the total number of patterns is just 32. Since the original dataset also displays missing values in some of the examples, patterns with missing values are removed from the data in our experiment, so as to keep the comparison of the different models as simple as possible. Clearly, this choice has also the effect of making the task even more difficult: given that the original data contains 5 corrupted items, the number of available patterns in our dataset is reduced by 15.6%. For these reasons, the dataset is particularly interesting for evaluating the behavior of HRFs in a relatively challenging domain. The results of the experiment are shown in Table 6.4.

Table 6.4 Recognition accuracy values measured (by 3-fold cross-validation) on the Lung-Cancer dataset for BN, DN ($k = 3$), HRF ($k = k^* = 4$), MRFs ($k = 3$), and NB

	Recognition Accuracy	
	Average	Standard Deviation
BN	59.25%	05.23
DN	48.14%	05.23
HRF	62.95%	10.47
MRF	55.55%	09.07
NB	48.14%	05.23

An important detail concerning the experiment is the following. Since the number of training patterns is very small, we reduce the complexity penalty employed in the MDL heuristic for learning the structure of BNs from $\frac{par(h)}{2} \log |\mathbf{D}|$ to $\frac{par(h)}{4} \log |\mathbf{D}|$. The reason is that the relative weight of the penalty becomes more and more important as the number of data items decreases, which means that, for particularly small datasets, the structure learning algorithm may deliver too sparsely connected networks. In fact, the described correction to the MDL heuristic worked well for the Lung-Cancer experiment, improving the behavior of both BNs and HRFs.

The best result on the Lung-Cancer dataset is achieved again by HRFs, followed by BNs, MRFs, DNs, and NB. The outcome of this experiment lends itself to the same interpretation as the previous experiment. One interesting point to note concerning the Lung-Cancer application is that for HRFs, the standard deviation is higher than for BNs. The reason for this is the following. Given the 3-fold cross-validation strategy, HRFs and BNs produce exactly the same decisions in two test sessions, whereas in the third test session HRFs classify correctly one more pattern than BNs. Due to the small size of the dataset, this improvement in prediction accuracy increases significantly the related standard deviation. Given the results collected both in a synthetic and in a real-world domain, the evidence supports the hypothesis that hybrid random fields, as compared to other probabilistic graphical models, are a reliable pattern classification technique.

6.3.4 Link Prediction

This section describes a way of applying probabilistic graphical models to link prediction tasks. Section 6.3.4.1 formalizes the way we use probabilistic graphical models for learning to predict links. Two applications to the task of predicting references in scientific papers are described in Section 6.3.4.2, while in Section 6.3.4.3 we use the models as recommender systems for a movie database.

6.3.4.1 Ranking Strategy

In general terms, our link-prediction system has to deal with a set of users of a database and a set of items contained in the database, where the items can be papers, movies, or virtually anything else. For each user, information is available concerning which items in the database have already been chosen by that user. Formally, the aim of the system is to compute, for each user, a scoring function measuring the expected interest of the user for each item in the database. The goal is to measure the interest in items that the user has not yet considered, so that they can be ranked according to their relevance for the next choice the user will make.

We denote the set of database users as $\mathcal{U} = \{u_1, \ldots, u_m\}$, and the set of database items as $\mathcal{O} = \{o_1, \ldots, o_n\}$. For each user u_i, we have a set $\mathcal{O}_i \subset \mathcal{O}$ of items such that $\mathcal{O}_i$ contains the items already chosen by u_i. The aim of the link-prediction system is to provide, for each user u_i, a scoring function $s_i(o_j)$, defined for each item o_j in the database, such that, if $s_i(o_j) > s_i(o_k)$ for $j \neq k$, then the predicted interest of u_i in o_j is higher than the predicted interest of u_i in o_k. Therefore, the scoring function $s_i(o_j)$ allows to rank objects in the database according to their expected interest to u_i.

In order to predict the interest a user u_i will have in object o_j, the kind of information we try to exploit is the conditional probability of choosing (i.e *linking to*) o_j given the set $\mathcal{O}_i$ of objects that u_i is known to have already chosen. If we define a way to estimate that conditional probability using each model, then we can simply use the value of that probability as the score assigned by each model to the object being ranked. In other words, if o_j is the object we want to rank for u_i once we know the elements of $\mathcal{O}_i$, then for each model we need to specify a way to compute the value of the function $s_i(o_j)$, defined as follows:

$$s_i(o_j) = P\left(L_i(o_j) = 1 \mid \bigwedge_{k \neq j} L_i(o_k)\right) \tag{6.2}$$

where $L_i(o_j)$ is a boolean function such that

$$L_i(o_j) = \begin{cases} 1 & \text{if } u_i \text{ links to } o_j \\ 0 & \text{otherwise} \end{cases} \tag{6.3}$$

First, let us explain how the NB classifier can be applied to link prediction and collaborative recommendation. If o_j is the object we need to rank for user u_i, then

$$P\big(L_i(o_j) = 1 \mid \bigwedge_{k \neq j} L_i(o_k)\big) = \frac{P\big(L_i(o_j) = 1\big) P\big(\bigwedge_{k \neq j} L_i(o_k) \mid L_i(o_j) = 1\big)}{P\big(\bigwedge_{k \neq j} L_i(o_k)\big)} \qquad (6.4)$$

where the stated equality is just an instance of Bayes' theorem. If we assume (based on the structure of the naive Bayes model) that $L_i(o_j)$ makes the value of each $L_i(o_k)$ independent of any $L_i(o_{k^*})$ such that $k \neq k^* \neq j$, then Equation 6.4 can be further simplified:

$$P\big(L_i(o_j) = 1 \mid \bigwedge_{k \neq j} L_i(o_k)\big) = \frac{P\big(L_i(o_j) = 1\big) \prod_{k \neq j} P\big(L_i(o_k) \mid L_i(o_j) = 1\big)}{P\big(\bigwedge_{k \neq j} L_i(o_k)\big)} \qquad (6.5)$$

Following the standard usage of NB for collaborative filtering, as endorsed e.g. in [218, 261], we drop the denominator from Equation 6.5. Therefore, if o_j is the object we want to rank for user u_i, we compute the following scoring function:

$$\begin{aligned}
s_i(o_j) &= \log\Big(P\big(L_i(o_j) = 1\big) \prod_{k \neq j} P\big(L_i(o_k) \mid L_i(o_j) = 1\big)\Big) \\
&= \log P\big(L_i(o_j) = 1\big) + \sum_{k \neq j} \log P\big(L_i(o_k) \mid L_i(o_j) = 1\big)
\end{aligned} \qquad (6.6)$$

In order to estimate the probabilities referred to in Equation 6.6, our dataset is formalized as a set of patterns $\{\mathbf{d}_1, \ldots, \mathbf{d}_m\}$, such that each $\mathbf{d}_i$ is a boolean vector $(L_i(o_1), \ldots, L_i(o_n))$ specifying which objects were chosen by u_i. In other words, for each user we construct a corresponding pattern whose dimension is the total number of objects contained in the database. Therefore, the resulting dataset will contain a number of patterns which is equal to the number of database users. Given such a dataset, absolute and conditional probabilities are estimated by computing relative frequencies and exploiting them in the way described in Section 2.3. This formalization of the data is also exploited for applying the other graphical models.

On the other hand, the way that Bayesian networks, dependency networks, Markov random fields, and hybrid random fields are applied to ranking is the following. Given the formalization of the dataset described above, we first learn a model containing n (boolean) random variables $X_1, \ldots, X_n$, where n is the number of objects contained in the database and each X_j corresponds to object o_j. Once the model has been learned, the notion of MB provides a conceptually straightforward (and computationally very efficient) way of computing the value specified in Equation 6.2:

$$\begin{aligned}
s_i(o_j) &= P\Big(L_i(o_j) = 1 \mid \bigwedge_{k \neq j} L_i(o_k)\Big) \\
&= P\big(X_j = 1 \mid x_{1_i}, \ldots, x_{j-1_i}, x_{j+1_i}, \ldots, x_{n_i}\big) \\
&= P\big(X_j = 1 \mid mb_i(X_j)\big)
\end{aligned} \qquad (6.7)$$

where each x_{k_i} is the value of $L_i(o_k)$, and $mb_i(X_j)$ is the state of the Markov blanket of X_j in the graphical model, as that state is determined by pattern $\mathbf{d}_i$.

6.3.4.2   Predicting References in Scientific Papers

The task we deal with in this section is the prediction of references in research papers. In particular, given a paper containing a specific set of references, the task is to rank all remaining papers in a certain database, based on their relevance as additional references to be included in the paper at hand.

We test the ranking algorithms on two different datasets, CiteSeer and Cora. We exploit a preliminary preprocessing of the data which is publicly available at http://www.cs.umd.edu/projects/linqs/projects/lbc/ index.html. From each dataset we extract the citation graph of the paper corpus. We then check the number of references contained in each paper, and we remove papers that do not contain at least 3 references. After this preprocessing, we formalize each dataset as a list containing m vectors of n boolean features, where each vector is a paper and each feature stands for the presence or absence of a certain reference within the paper. For the CiteSeer dataset we have that $m = 547$ and $n = 1,067$, while for Cora we have that $m = 956$ and $n = 1,229$. In other words, based on the formalism described in Section 6.3.4.1, m corresponds to the number of database users, while n corresponds to the number of database objects.

Each dataset is partitioned into training and test sets according to a 5-fold cross-validation procedure. After training, the test consists in the following task. We remove one reference from each paper in the test set, and we require the tested model to rank that reference given the remaining references contained in the paper at hand. The idea behind this query is that the removed reference should receive the highest possible rank from a good ranking algorithm, since that reference is probably among the most relevant ones for the paper at hand.

The results of ranking are evaluated using the mean reciprocal rank (MRR) and success rate (SR) at N metrics, which are widely employed accuracy measures in information retrieval research. If j is the index of a certain example (i.e. of a given query) within the test set, o_j is the object (i.e. the reference) that should receive the highest rank for that query, and $rank(o_j)$ is the rank assigned to o_j by the algorithm at hand, then MRR is defined as follows:

$$MRR = \frac{1}{m} \sum_{j=1}^{m} \frac{1}{rank(o_j)} \qquad (6.8)$$

where m is the size of the test set. SR is defined instead with respect to a parameter N, so as to measure the capability of an algorithm to rank the target object o_j within the top N candidates. Given the Heaviside function $\Theta(x)$, such that $\Theta(x) = 1$ if $x \geq 0$ and $\Theta(x) = 0$ otherwise, SR at N is defined in the following way:

$$SR(N) = \frac{1}{m} \sum_{j=1}^{m} \Theta(N - rank(o_j)) \tag{6.9}$$

The results of the application to the CiteSeer dataset are shown in Table 6.5 and Figure 6.6, while Table 6.6 and Figure 6.7 provide results for the Cora dataset. All values are measured based on a 5-fold cross-validation procedure.

An important remark concerns the way we train the standard BN model in the CiteSeer and Cora experiments. The dimensionality of these tasks prevents us from applying Algorithm 2.1 successfully, because of computational limitations. On the CiteSeer dataset (that is for a problem involving 1,067 network nodes), 72 hours do not suffice for a 1.83 GHz CPU to return the output of Algorithm 2.1. Therefore, we train the BN model using the K2 structure learning algorithm, where the maximum number of parents allowed for each node is set to 2. Although the K2 algorithm is much faster than Algorithm 2.1 (at the cost of being less accurate), it is still very expensive to run for high-dimensional problems. In fact, its worst-case computational complexity is $O(n^4)$, as shown by [50]. On the CiteSeer dataset, K2 requires about 20 hours computation (on the same 1.83 GHz PC architecture), while for Cora (that is for 1,229 network nodes) it requires about 40 hours. Of course, all time measurements concern training for one fold only, and not the whole execution of 5-fold cross-validation. On the other hand, HRFs do not suffer at all from computational problems. The difference between running MBM on the CiteSeer dataset and running it on Cora is relatively small (in the order of a few minutes), and in both cases training time does not exceed 1 hour. Training time for NB is also not significant (about half an hour on Cora), whereas MRFs require about 6 hours training for CiteSeer and 8 hours for Cora. Finally, learning DNs only takes between 10 and 20 minutes.

Table 6.5 MRR measured (by 5-fold cross-validation) on the CiteSeer dataset for BN, DN ($k = 8$), HRF ($k = 8$, $k^* = 10$), MRF ($k = 3$), and NB

	Mean Reciprocal Rank	
	Average	Standard Deviation
BN	0.2823	0.0470
DN	0.0130	0.0010
HRF	0.2865	0.0251
MRF	0.1771	0.0306
NB	0.0537	0.0108

The results of the experiments are quite encouraging for hybrid random fields. Both HRFs and BNs significantly outperform NB, DNs, and MRFs. While HRFs achieve better average results than BNs on CiteSeer, especially with respect to the $SR(N)$ metric, on the Cora dataset BNs are more accurate than HRFs with respect to the MRR metric, although the two models are nearly equivalent in that case with respect to SR. Furthermore, in both tasks the standard deviation of MRR values is

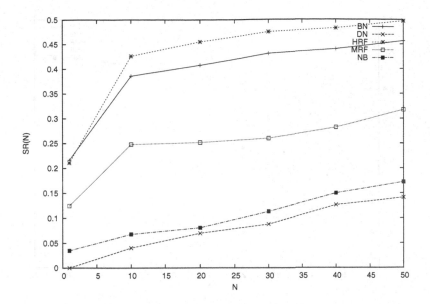

Fig. 6.6 Average SR at N measured on the CiteSeer dataset for BN, DN ($k = 8$), HRF ($k = 8$, $k^* = 10$), MRF ($k = 3$), and NB. Values are measured by 5-fold cross-validation

Table 6.6 Average MRR measured (by 5-fold cross-validation) on the Cora dataset for BN, DN ($k = 8$), MRF ($k = 3$), HRF ($k = 8$, $k^* = 10$), and NB

	Mean Reciprocal Rank	
	Average	Standard Deviation
BN	0.2916	0.0228
DN	0.0417	0.0034
MRF	0.0704	0.0096
HRF	0.2647	0.0168
NB	0.1519	0.0131

lower for HRFs than it is for BNs. In general, the behavior of HRFs and BNs is much more reliable across the two tasks than the latter models, for the following reason. While MRFs outperform NB on the CiteSeer dataset, NB outperforms instead MRFs the Cora dataset. Given this result, the kind of probability distribution underlying the former domain must be significantly different from the distribution displayed by the second one. Now, since both HRFs and BNs behave stably well across the two tasks, the experiments show these two models to be more robust than the competing ones, at least in the considered range of applications.

Note that, in our experimental setting, the ranking algorithms do not take into account the publication year of the papers they are trained/tested on. Therefore, citations are sometimes recommended (during the test sessions) that cannot possibly

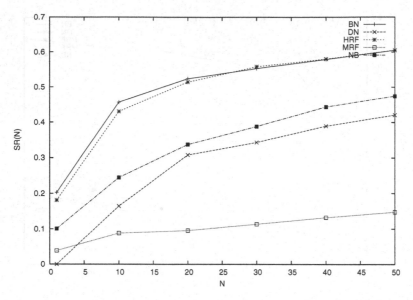

Fig. 6.7 Average SR at N measured (by 5-fold cross-validation) on the Cora dataset for BN, DN ($k = 8$), HRF ($k = 8$, $k^* = 10$), MRF ($k = 3$), and NB

lead to correct predictions because of chronological impossibility. Since ignoring the publication year of papers (when recommending citations) lowers the measured accuracy for all compared algorithms, this feature of the experiment does not affect the significance of the comparison.

6.3.4.3 Predicting Preferences for Movies

The task we deal with in this section involves the MovieLens database. The dataset used in the experiment contains data concerning 1,682 movies. The number of database users is 943. The MovieLens dataset is publicly available at http://www.grouplens.org/. We formalize the MovieLens link-prediction task according to an *implicit-voting-with-binary-preferences* strategy [33]. 'Implicit voting' means that we only exploit information concerning whether a user rated a certain item or not, without taking into account the specific rating. Therefore, user choices are modeled as 'binary preferences'. Given such a formalization, applying the ranking strategy described in Section 6.3.4.1 is straightforward.

Results on the MovieLens link-prediction task are evaluated using two versions of the degree of agreement (DOA) metric, which we describe next. The DOA metric is aimed at measuring how accurate a ranking of database items is for a user u_i. Let us denote by $\mathcal{L}$ a given training set and by $\mathcal{T}$ the corresponding test set. Moreover, let $\mathcal{L}_i$ denote the set of movies such that each one of these movies is rated by u_i in $\mathcal{L}$, and $\mathcal{T}_i$ the set of movies rated by u_i in $\mathcal{T}$. Finally, let $\mathcal{N}_i$ denote the set of movies that are never rated by u_i, so that $\mathcal{N}_i = \mathcal{O} \setminus (\mathcal{L}_i \cup \mathcal{T}_i)$. The first step in specifying the DOA for a ranking algorithm R is to define, for each user u_i and for any pair of movies o_j and o_k, a function $order_i(o_j, o_k)$ such that

$$order_i(o_j, o_k) = \begin{cases} 1 & \text{if } R_i(o_j) > R_i(o_k) \\ 0 & \text{otherwise} \end{cases} \tag{6.10}$$

where $R_i(o)$ is the rank assigned by R to movie o for user u_i. Given this function, we define the DOA for each user u_i in the following way:

$$DOA_i = \frac{\sum_{o_j \in \mathcal{T}_i \wedge o_k \in \mathcal{N}_i} order_i(o_j, o_k)}{|\mathcal{T}_i| \cdot |\mathcal{N}_i|} \tag{6.11}$$

For each user u_i, DOA_i measures the percentage of movie pairs in $\mathcal{T}_i \times \mathcal{N}_i$ ranked in the correct order by the algorithm at hand. Once defined the DOA with respect to each user u_i, we specify *macro-averaged* and *micro-averaged* DOA. The macro-averaged DOA is defined as

$$\text{macro-DOA} = \frac{\sum_{\mathcal{T}_i \neq \emptyset} DOA_i}{|\{\mathcal{T}_j : \mathcal{T}_j \neq \emptyset\}|} \tag{6.12}$$

On the other hand, the micro-averaged DOA is given by

$$\text{micro-DOA} = \frac{\sum_{\mathcal{T}_i \neq \emptyset} \sum_{o_j \in \mathcal{T}_i \wedge o_k \in \mathcal{N}_i} order_i(o_j, o_k)}{\sum_{\mathcal{T}_l \neq \emptyset} |\mathcal{T}_l| \cdot |\mathcal{N}_l|} \tag{6.13}$$

Clearly, the micro-DOA assigns a higher weight to users with a larger number of ratings in the test set, whereas the macro-DOA assigns the same weight to all users, no matter how many ratings are present in the test set for each one of them.

Macro-averaged and micro-averaged DOA values are measured by 5-fold cross-validation. The test employs a publicly available partitioning of the dataset, which allows to easily compare different results to be found in the literature. Training the MRF model requires about 12 hours on average for each fold (on the same PC architecture used for the previous measurements). The number of feature functions contained in each MRF (averaged over the five models learned for the different folds) is equal to 16,396. This means that, in order to learn the weights of each MRF, we need to optimize a function of 16,396 parameters (on average). On the other hand, learning HRFs by MBM takes about one hour and a half for each fold, while training DNs requires less than one hour.

For this application, the prediction accuracy of graphical models is also compared to the accuracy achieved by the ItemRank (IR) recommender system [117], which is known to perform pretty well on the MovieLens database with respect to the used evaluation metric. IR is a biased version of the PageRank algorithm [36], modified in order to be applied to recommender systems. The first step in IR is to construct a correlation matrix $\mathbf{C}$ such that each element $\mathbf{C}_{i,j}$ is the number of users who rated both object o_i and object o_j. The columns in $\mathbf{C}$ are then normalized so as to obtain a stochastic matrix. The second step is to construct and normalize, for each user u_i, a vector $\mathbf{r}_i$ of size $|\mathcal{O}|$ such that each element r_j^i of $\mathbf{r}_i$ is the (normalized) rating expressed by u_i for object o_j. Given the matrix $\mathbf{C}$ and the vector $\mathbf{r}_i$, a ranking for the movies in the database is represented as a vector $\mathbf{IR}_i$ such that each vector element IR_j^i is the score assigned by IR to movie m_j for user u_i. In order to compute the values of $\mathbf{IR}_i$, the following system of equations has to be run recursively until convergence:

$$\mathbf{IR}_i(t) = \begin{cases} \frac{1}{|\mathcal{O}|} \cdot \mathbf{1}_{|\mathcal{O}|} & \text{if } t = 0 \\ \alpha \cdot \mathbf{C} \cdot \mathbf{IR}_i(t-1) + (1-\alpha) \cdot \mathbf{r}_i & \text{if } t > 0 \end{cases} \tag{6.14}$$

where $\mathbf{1}_{|\mathcal{O}|}$ denotes a vector of size $|\mathcal{O}|$ such that each vector element is 1, and α is a constant (typically set to 0.85). Clearly, the vector $\mathbf{r}_i$ serves the purpose of biasing the computation so as to reflect the preferences of u_i.

Tables 6.7 and 6.8 collect the results of the experiment, as measured by the macro-averaged and micro-averaged DOA respectively. Bayesian networks are not included in the comparison because the K2 algorithm does not finish running in 72 hours.

Table 6.7 Macro-DOA measured (by 5-fold cross-validation) on the MovieLens dataset for DN ($k = 8$), HRF ($k = 8$, $k^* = 10$), IR, MRF ($k = 3$), and NB

	Macro-averaged DOA	
	Average	**Standard Deviation**
DN	0.8051	0.0123
HRF	0.8983	0.0052
IR	0.8776	0.0027
MRF	0.8947	0.0044
NB	0.8887	0.0022

Table 6.8 Micro-DOA measured (by 5-fold cross-validation) on the MovieLens dataset for DN ($k = 8$), HRF ($k = 8$, $k^* = 10$), IR, MRF ($k = 3$), and NB

	Micro-averaged DOA	
	Average	**Standard Deviation**
DN	0.8133	0.0043
HRF	0.8807	0.0059
IR	0.8706	0.0010
MRF	0.8809	0.0050
NB	0.8666	0.0030

Hybrid random fields achieve the highest accuracy with respect to the macro-averaged DOA, while its performance is nearly equivalent to the best one (achieved by Markov networks) with respect to the micro-averaged DOA. Both MRFs and HRFs are more accurate than NB and IR, according to both evaluation metrics. On the other hand, the accuracy displayed by DNs is significantly lower with respect to all other models.

6.4 Pattern Classification in Continuous Domains

The aim of this section is to evaluate the accuracy of nonparametric HRFs at modeling (multivariate) densities featuring nonlinear dependencies between the variables plus random noise (distributed in heterogeneous ways). To this aim, we sample a number of datasets from synthetic distributions, where the distributions are randomly generated in such a way as to make it unlikely that any particular parametric assumption may be satisfied. We then exploit the produced data for pattern classification, comparing the performance of our model to other probabilistic techniques. The data generation process is described in Section 6.4.1, while Section 6.4.2 illustrates the results of the experiments.

6.4.1 Random Data Generation

In order to generate datasets featuring nonlinear correlations between the variables, we exploit the idea of defining a random distribution based on a (randomly generated) DAG, where each node corresponds to a random variable and each arc corresponds to a dependence of the child on the parent. Therefore, the data generation process is made up of three stages: first, we generate a random DAG with a specified number of nodes; second, we generate a random distribution from a specified DAG; third, we generate a random dataset from a specified (DAG-shaped) distribution.

6.4.1.1 Directed Acyclic Graph Generation

Given a number d of nodes and a parameter p_{max} specifying the maximum number of parents allowed for each node, we generate a random DAG using Algorithm 6.1. We start by assuming a random ordering of the nodes from X_1 to X_n. Then, for each X_i, we randomly select p nodes from the set $\{X_j : j < i\}$, where p is a random integer in the interval $[0, \min\{i - 1, p^*\}]$, and for each selected node X_j we introduce an edge from X_j to X_i. The resulting pair $(\mathcal{V}, \mathcal{E})$, where $\mathcal{V}$ is the set of vertices and $\mathcal{E}$ is the set of edges, is returned as output.

Algorithm 6.1 `generateRandomDAG`: Random DAG generation

Input: Integers d, p^*.
Output: DAG $\mathcal{G} = (\mathcal{V}, \mathcal{E})$.

```
generateRandomDAG(d, p^max):
1.     V = {X_1,...,X_d}
2.     E = ∅
3.     for(i = 1 to d)
4.         p = random integer in [0,min{i-1,p*}]
5.         P = ∅
6.         while(|P| < p)
7.             j = random integer in [1,i-1]
8.             P = P∪{(X_j,X_i)}
9.         E = E∪P
10.    return (V,E)
```

6.4.1.2 Distribution Generation

Algorithm 6.2 generates a random distribution from a DAG $\mathcal{G} = (\mathcal{V}, \mathcal{E})$. The idea is that each edge (X_i, X_j) in the DAG represents a dependence of X_j on X_i, where the dependence is determined by a polynomial function of third degree $f_{ji}(\cdot)$, defined as

$$f_{ji}(x) = a_1^{ji} x^3 + a_2^{ji} x^2 + a_3^{ji} x + a_4^{ji} \qquad (6.15)$$

The coefficients $a_1, \ldots, a_4$ of each polynomial are selected randomly in the interval $[-a^*, a^*]$. Moreover, each node X_i is assigned a beta density function $beta_i(\cdot)$, defined as follows (for $a < x < b$ and $\alpha_i, \beta_i > 0$):

$$beta_i(x) = \frac{\Gamma(\alpha_i + \beta_i)}{\Gamma(\alpha_i) \Gamma(\beta_i) (b-a)^{\alpha_i + \beta_i - 1}} (x-a)^{\alpha_i - 1} (b-x)^{\beta_i - 1} \qquad (6.16)$$

where

$$\Gamma(x) = \int_0^\infty t^{x-1} e^{-t} \, dt \qquad (6.17)$$

The idea is that the values observed for variable X_i are subject to random noise, where the noise is distributed over the interval (a, b) according to a beta density with parameters α_i and β_i. For each $beta_i$, the parameters α_i and β_i are randomly chosen in the intervals $(0, \alpha^*]$ and $(0, \beta^*]$ respectively, whereas a and b remain constant. Given the polynomials and the beta densities, the value of each X_i results from a linear combination of the related polynomial functions plus (beta-distributed) random noise. The output of Algorithm 6.2 is a pair $(\mathcal{F}_\mathcal{V}, \mathcal{F}_\mathcal{E})$ such that $\mathcal{F}_\mathcal{V} = \{beta_i : X_i \in \mathcal{V}\}$ and $\mathcal{F}_\mathcal{E} = \{f_{i_j} : (X_j, X_i) \in \mathcal{E}\}$.

Algorithm 6.2 PDBNGenerateRandomDistribution: Generating a random distribution with DAG-shaped polynomial dependencies and beta-distributed noise

---

Input: DAG $\mathcal{G} = (\mathcal{V}, \mathcal{E})$; positive real numbers α^*, β^*, a^*; real numbers a, b such that $a < b$.
Output: DAG-shaped distribution $p^{\mathcal{G}} = (\mathcal{F}_{\mathcal{V}}, \mathcal{F}_{\mathcal{E}})$.

```
PDBNGenerateRandomDistribution(𝒢,α*,β*,a*,a,b):
 1.    ℱ𝒱 = ∅
 2.    for(i = 1 to |𝒱|)
 3.        αᵢ = random real in (0,α*]
 4.        βᵢ = random real in (0,β*]
 5.        betaᵢ(·) = beta(·;αᵢ,βᵢ,a,b)  //See Equation 6.16
 6.        ℱ𝒱 = ℱ𝒱∪{betaᵢ}
 7.    ℱℰ = ∅
 8.    for((Xᵢ,Xⱼ) ∈ ℰ)
 9.        for(k = 1 to 4)
10.            aₖʲⁱ = random real in [−a*,a*]
11.        fⱼᵢ(·) = f(·;a₁ʲⁱ,...,a₄ʲⁱ)  //See Equation 6.15
12.        ℱℰ = ℱℰ∪{fⱼᵢ}
13.    return (ℱ𝒱,ℱℰ)
```

6.4.1.3 Dataset Generation

Given a distribution $p^{\mathcal{G}} = (\mathcal{F}_{\mathcal{V}}, \mathcal{F}_{\mathcal{E}})$ organized according to DAG $\mathcal{G} = (\mathcal{V}, \mathcal{E})$, Algorithm 6.3 generates patterns that are independent and identically distributed according to $p^{\mathcal{G}}$. In order to produce a pattern $x_1, \ldots, x_d$, the algorithm determines the value of each variable X_i by first computing $\sum_{f_{ij}(x) \in \mathcal{F}_{\mathcal{E}}} f_{ij}(x_j)$, and then by adding to that sum a random value sampled from the density $beta_i(x)$, so as to introduce some noise. The ancestral ordering of the nodes $X_1, \ldots, X_d$ in $\mathcal{V}$ is followed so as to ensure that the argument of each function $f_{ij}(x_j)$ has already been determined before computing the value of node X_i.

If one needs to generate data that are partitioned into several classes $\omega_1, \ldots, \omega_c$ (e.g. for the purposes of pattern classification), the algorithm generates data for each ω_i such that $i > 1$ by deriving first a corresponding distribution $p_i^{\mathcal{G}}$ from $p_{i-1}^{\mathcal{G}}$ in the following way. For each polynomial $f_{j_k}(x)$ in $\mathcal{F}_{\mathcal{E} i-1}$, the coefficients $a_1^{j_k}, \ldots, a_4^{j_k}$ are changed with probability P, where the change consists in multiplying each $a_l^{j_k}$ by a randomly selected real number in the interval $[-r^*, r^*]$. The resulting polynomial is used to replace $f_{j_k}(x)$ in $\mathcal{F}_{\mathcal{E} i}$. Finally, the integers $n_1, \ldots, n_c$ specify the number of patterns to be generated for each class.

Algorithm 6.3   `PDBNGenerateRandomData`: Generating random data from a DAG-shaped distribution

Input: DAG-shaped distribution $p^{\mathcal{G}} = (\mathcal{F}_V, \mathcal{F}_{\mathcal{E}})$; number c of classes; set of integers $\mathbf{n} = \{n_1, \ldots, n_c\}$; real numbers P, r^* such that $0 < P \le 1, r^* > 0$.
Output: Datasets $\mathbf{D}_1, \ldots, \mathbf{D}_c$.

```
PDBNGenerateRandomData(pᴳ,c,n,P,r*):
 1.    𝒟 = {D₁,...,D_c}
 2.    for(i = 1 to c)
 3.        Dᵢ = ∅
 4.        if(i > 1)
 5.            distributionIsUnchanged = true
 6.            while(distributionIsUnchanged)
 7.                for(f_{jₖ}(x) ∈ 𝓕_ℰ)
 8.                    p = random real in [0,1)
 9.                    if(p < P)
10.                        for(l = 1 to 4)
11.                            r = random real in [−r*,r*]
12.                            a_l^{jk} = r a_l^{jk}
13.                            f_{jₖ}(·) = f(·;a₁^{jk},...,a₄^{jk})  //See Equation 6.15
14.                            distributionIsUnchanged = false
15.        for(j = 1 to nᵢ)
16.            for(k = 1 to d)
17.                x_{kⱼ} = random real sampled from betaₖ
18.                x_{kⱼ} = x_{kⱼ} + Σ_{f_{kₗ(x)}∈𝓕_ℰ} f_{kl}(x_{lⱼ})
19.            Dᵢ = Dᵢ∪{(x₁ⱼ,...,x_{dⱼ})}
20.    return 𝒟
```

6.4.2   Results

In order to test the accuracy of kernel-based HRFs (KHRFs) at modeling joint densities (as learned by MBM), we apply them to a number of pattern classification tasks, where the datasets are generated using Algorithms 6.1–6.3. We consider eleven tasks, where each task is based on a different dataset $\mathbf{D}$ containing 500 patterns, and the patterns are equally divided in two classes ω_1 and ω_2. The data for each task are generated using each time a different (random) DAG. In particular, we choose a different number d of nodes for each DAG, where $10 \le d \le 20$. Then, we use the generated DAG as input for Algorithm 6.2. Here, we set $a^* = 2$ for the polynomial functions, while the beta densities are generated over the interval $[-2,2]$, setting $\alpha^* = \beta^* = 2$. In a preliminary phase of the experiments, we found these parameters to be large enough to generate a suitably wide range of distributions. We use $c = 2$ and $n_1 = n_2 = 250$ as input values for Algorithm 6.3. Moreover, when changing the distribution from ω_1 to ω_2, we set $r^* = 2$ and $P = 0.1$. Our experience with preliminary results indicated that if the values of r^* and P (especially the latter) are too large (e.g. if $P \gtrsim 0.2$), the resulting classification tasks tend to be too easy to be dealt with, because patterns belonging to different classes are then distributed farther apart in the feature space. Before exploiting the datasets, we normalize the values of each feature X_i by transforming each x_{i_j} into $\frac{x_{i_j} - \min_k x_{i_k}}{\max_k x_{i_k} - \min_k x_{i_k}}$, where $1 \leq k \leq |\mathbf{D}|$.

In order to give a concrete idea of the kind of distributions generated using the method and parameter settings described above, some examples are plotted in Figure 6.8, where the involved distributions are organized in a DAG $\mathcal{G} = (\mathcal{V}, \mathcal{E})$ such that $\mathcal{V} = \{X_1, X_2\}$ and $\mathcal{E} = \{(X_1, X_2)\}$. For each plot, the x-axis corresponds to the X_1 node of the DAG, while the y-axis corresponds to the X_2 variable. Notice how the random choice of the parameter values (both for the dependence relationship and the noise function) is flexible enough to generate a relatively wide range of cases. For instance, the $beta_2$ function (i.e. the density function associated with variable X_2) produces nearly uniform density (over the support of the distribution) for Figure 6.8a, while it generates noise which is peaked toward the lower/higher extreme in Figure 6.8b/6.8c, or toward both extremes in Figure 6.8d. On the other hand, the parameters of the polynomial function are able to determine nearly-quadratic dependences (Figures 6.8a, 6.8b), cubic dependences (Figure 6.8c), and nearly-linear dependences (class 2 in Figures 6.8d–6.8f). Generally, the employed data generation technique is capable of producing a relatively wide variety of pattern classification problems: 'easy' problems, where the classes are linearly separable (such as in Figure 6.8f); moderately difficult tasks, where the classes may overlap to a significant extent but a linear separation might be settled for with relatively good results (such as in Figure 6.8a); fairly hard problems, where patterns drawn from different classes are neither linearly separable nor belonging to neatly separated regions of the feature space (such as in Figures 6.8b–6.8e).

We compare the performance of KHRFs to kernel-based BNs (KBNs), kernel-based MRFs (KMRFs), nonparanormal Markov random fields (NPMRFs), and Gaussian MRFs (GMRFs). In kernel-based BNs and MRFs we estimate conditional densities in the same way as in KHRFs, while the model structure is learned using the algorithms proposed in [136] and [137] respectively. For the purposes of bandwidth selection, we always perform two iterations of Algorithm 5.1, whereas the limit points h_{max_1} and h_{max_2} are set differently for BNs, MRFs, and HRFs, based on preliminary validation on separate datasets. In particular, the used values are $h_{max_1} = 2$ and $h_{max_2} = 1$ for KBNs, $h_{max_1} = 1$ and $h_{max_2} = 2$ for KMRFs, and $h_{max_1} = 0.05$ and $h_{max_2} = 0.5$ for KHRFs. Structure learning in GMRFs and NPM-RFs is performed as described in [98] and [197], using the graphical lasso technique, and conditional densities are then estimated within the resulting structures using Gaussian and nonparanormal conditional density models respectively. To the best of our knowledge, the learning algorithms considered for KBNs, KMRFs, and NPMRFs are the state of the art in the literature on (continuous) nonparametric and semiparametric graphical models. On the other hand, GMRFs provide an authoritative term of comparison for evaluating the effect of relaxing the parametric assumption in density estimation.

In order to exploit the models for pattern classification, we use the maximum a posteriori strategy (based on Bayesian decision theory) described in Section 6.3.3.

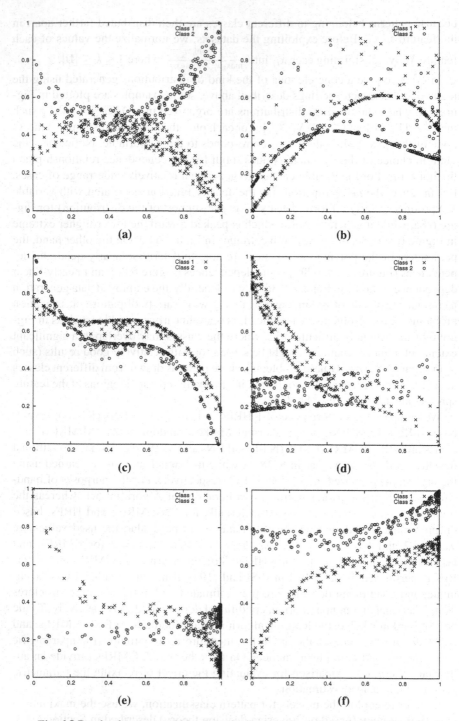

Fig. 6.8 Randomly generated bivariate distributions for pattern classification tasks

The results of the experiments are reported in Tables 6.9–6.10, where all values are measured by 5-fold cross-validation. For each model, we report both recognition accuracy and training time (per class), where time was measured (in seconds) on a 2.34 GHz CPU.

Tables 6.9–6.10 lend themselves to the following interpretation. First, KHRFs are more accurate overall than the other models in terms of recognition rate. At the same time, learning KHRFs is much less expensive than learning kernel-based BNs and MRFs. Second, although GMRFs and NPMRFs are the most efficient models from the computational point of view, their advantage over KHRFs against the

Table 6.9 Recognition accuracy (average ± standard deviation) measured by 5-fold cross-validation on synthetic datasets of growing dimensionality. For each dataset, d is the number of variables composing the data vectors

d	Recognition Accuracy (%)				
	KBN	GMRF	NPMRF	KMRF	KHRF
10	59.4 ± 4.4	61.2 ± 2.0	61.4 ± 03.8	59.6 ± 4.5	62.6 ± 7.6
11	48.8 ± 4.6	51.6 ± 2.0	54.8 ± 04.5	49.2 ± 5.2	62.4 ± 4.8
12	80.8 ± 2.8	70.6 ± 4.2	87.2 ± 03.1	79.0 ± 2.6	82.2 ± 2.3
13	56.2 ± 3.9	53.2 ± 2.4	56.4 ± 33.6	56.4 ± 4.6	65.0 ± 4.5
14	83.8 ± 4.7	87.2 ± 2.3	75.6 ± 07.7	85.0 ± 5.0	88.4 ± 2.9
15	68.0 ± 2.6	58.8 ± 3.4	30.0 ± 24.4	68.4 ± 4.8	73.8 ± 3.7
16	55.0 ± 6.3	76.0 ± 4.6	66.4 ± 07.3	63.2 ± 4.5	81.4 ± 1.7
17	52.2 ± 2.1	50.6 ± 1.3	63.4 ± 01.8	55.8 ± 3.9	56.4 ± 2.4
18	58.2 ± 7.3	53.6 ± 2.7	64.0 ± 03.2	62.2 ± 5.9	75.2 ± 2.7
19	97.8 ± 0.7	97.6 ± 1.8	91.0 ± 01.7	98.4 ± 0.8	98.8 ± 0.7
20	96.6 ± 2.5	98.4 ± 1.0	78.8 ± 03.5	96.6 ± 1.4	97.4 ± 1.2

Table 6.10 Average training time (per class) measured by 5-fold cross-validation on synthetic datasets of growing dimensionality. For each dataset, d is the number of variables composing the data vectors.

d	Training Time (s)				
	KBN	GMRF	NPMRF	KMRF	KHRF
10	031.5	0.2	0.2	025.7	3.0
11	033.8	0.2	0.2	031.9	4.7
12	057.6	0.2	0.2	027.1	4.7
13	063.9	0.2	0.2	060.3	3.1
14	093.0	0.2	0.3	069.2	6.1
15	090.4	0.2	0.3	093.7	5.0
16	146.7	0.2	0.3	082.8	3.0
17	141.3	0.2	0.3	102.8	3.4
18	152.1	0.3	0.3	136.7	3.9
19	196.2	0.3	0.3	167.2	4.3
20	270.0	0.3	0.3	212.4	6.0

growth of the number of variables is not as significant as the advantage of GMRFs, NPMRFs, and KHRFs over KBNs and KMRFs. Third, the relatively low accuracy of GMRFs as compared to KHRFs, together with the fact that the improvement of NPMRFs over GMRFs is not as stable as one may wish, confirms that the distributions generating the datasets violate the parametric and semiparametric assumptions to a significant extent. Therefore, the considered experimental setting provides evidence not only that kernel-based HRFs are a very reasonable choice when no prior knowledge is available concerning the form of the distribution to be estimated, but also that KHRFs are the most promising option within the kernel-based family, both in terms of computational efficiency and prediction accuracy.

6.5   Final Remarks

We finally put hybrid random fields at work, with very promising results. This chapter gave a demonstration of how probabilistic graphical models can be used for feature selection—exploiting their structural capability of capturing dependencies among variables that are actually relevant to pattern recognition tasks—in order to reduce the dimensionality of the feature vectors to be passed on to a traditional classifier. We also pointed out how the graphical models, intrinsically encapsulating a joint probability distribution, can be effectively applied to classification tasks (which, indeed, require modeling posterior class probabilities). Experiments on a number of classification and link prediction datasets, over discrete and continuous variables, provided us with significant empirical evidence of (*i*) the quality of the probability/density estimation capabilities offered by HRFs, which compares favorably with the established models, (*ii*) the viability of HRFs as solutions to classification and prediction tasks, and (*iii*) the scalability of the proposed learning algorithms, whose computational burden proves to be pretty much tractable even for a significantly growing dimensionality of the datasets. So far, throughout the last three chapters, we stated the formal definitions of the new paradigm, we established its formal properties, we presented suitable algorithms for inference and learning (over both discrete and continuous variables), and we accomplished the empirical evaluation. Now, we can finally turn our attention to some broader questions and try to get 'the big picture' by putting the developed models (and the general methodology employed for the overall research) into the proper philosophical frame. The next chapter pictures probabilistic graphical models and statistical machine learning from a cognitive-science point of view.

Chapter 7
Probabilistic Graphical Models: Cognitive Science or Cognitive Technology?

"We believe that performing experiments with different discovery methods will provide useful information about the intrinsic difficulty of discovery, as well as about the power of particular search heuristics."

Pat Langley, Herbert A. Simon, Gary L. Bradshaw,
and Jan M. Zytkow, 1987 [187]

7.1 Introduction

This chapter is an attempt to make explicit the philosophical and cognitive perspective that the scientific work presented in Chapters 2–6 should be viewed from. This does not mean that the scientific material collected in this work needs a philosophical foundation in order to make sense or to be really interesting. The only aim of embedding scientific results within a philosophical framework is "to understand how things in the broadest possible sense of the term hang together in the broadest possible sense of the term" [275], which is what Wilfrid Sellars regarded as the general aim of philosophy. In other words, while proposing a philosophical reflection on the meaning of the technical results collected in the previous chapters, we do not think that the value of those results depends in any important way on their philosophical meaning. Our standpoint is rather that, *if* we ask how those results in AI "hang together" with other results in the cognitive sciences and with particular views advocated in the philosophy of mind, *then* the philosophical remarks contained in this chapter are the answer we give to that question. But the reader should keep in mind that our 'philosophical' reflections are more properly meant as a scientific contribution to philosophy, rather than a philosophical contribution to science, where the guiding idea is that science can take care of itself.

In Section 7.2, we try to get rid of a misleading (and somewhat dangerous) psychologistic conception of the role of AI within the cognitive sciences. To that aim, we relate that misconception of AI to a simplistic and untenable interpretation of Turing's seminal reflection on machine intelligence. Section 7.3 replaces the view of AI as one of the cognitive sciences with a view of AI as cognitive technology, i.e. as a formal and experimental investigation of computational techniques for advancing natural cognition. AI is not committed at all to viewing natural cognition as a computational process, but it designs and investigates artificial computational processes for aiding, complementing, and extending natural cognition. Finally, Section 7.4 will point at the capability of machine learning research to offer new

A. Freno and E. Trentin: Hybrid Random Fields, ISRL 15, pp. 151–162.
springerlink.com   © Springer-Verlag Berlin Heidelberg 2011

perspectives and insights on traditional issues in the theory of knowledge: while the philosophical views advocated in this chapter are somewhat deflationary concerning the relevance of AI to the philosophy of mind, they stress instead the relevance of machine learning to some problems in epistemology [179, 291, 292].

7.2   A Philosophical View of Artificial Intelligence

The general view of artificial intelligence we are going to argue for in this section may be referred to as an 'eliminative philosophy' of AI. While the name of the views advocated in this section is clearly borrowed from Paul Churchland's views in the philosophy of mind [48], the reader should not assume in advance any kind of similarity between eliminativism in the philosophy of AI and eliminative materialism as a theory of mind. By eliminativism in the philosophy of AI, we mean the view that *the set of concepts, methods, and principles employed in current AI research does not bear any necessary relationship to the set of concepts, methods, and principles underlying the study of natural minds* (as carried out e.g. by cognitive psychology or by cognitive neuroscience). In other words, our view maintains that the principles of natural cognition, as they are uncovered by the different cognitive sciences, do not—and should not—impose any constraint on the way that intelligent machines are built and investigated by artificial intelligence researchers. Of course, the philosophy of AI we are introducing is not meant to deny either that knowledge of how natural minds work can contribute in important ways to the progress of AI, or that AI achievements can contribute to the understanding of natural cognition. The point of eliminativism is simply to deny that any connection between AI and the study of natural intelligence is a *necessary* connection, i.e. a 'constitutive property' of AI as a scientific discipline. In order to justify such a view of the relationship between AI and the other cognitive sciences, we put forward two different arguments: the first argument is presented in Section 7.2.1, while the second one is offered in Section 7.2.2.

7.2.1   The Argument from Authority

The first argument for eliminativism in the philosophy of AI, which we call 'the argument from authority', is mainly based on some of Alan Turing's views, as expressed in the seminal paper "Computing Machinery and Intelligence" [306]. The reason for the argument's name will become clear as we will go through its details.

It is worth quoting entirely the first paragraph of Turing's paper on machine intelligence:

> I propose to consider the question, 'Can machines think?' This should begin with definitions of the meaning of the terms 'machines' and 'think'. The definitions might be framed so as to reflect so far as possible the normal use of the words, but this attitude is dangerous. If the meaning of the words 'machine' and 'think' are to be found by examining how they are commonly used it is difficult to escape the conclusion that the meaning and the answer to the question, 'Can machines think?' is to be sought in a

statistical survey such as a Gallup poll. But this is absurd. Instead of attempting such a definition I shall replace the question by another, which is closely related to it and is expressed in relatively unambiguous words. [306, p. 433]

Before commenting on this quote, let us briefly summarize how Turing replaces the initial, ambiguous question by a more precise one. The alternative question used by Turing in place of 'Can machines think?' is related to the outcome of an imitation game, which nowadays is widely known (although in a simplified version) as *Turing's test*. It is worth noting that the imitation game is not, in itself, the main concern of the argument from authority, i.e. the plausibility of the argument does not depend on the actual interest or usefulness of Turing's test. For the purposes of the argument, it suffices to understand that the imitation game provides a means of measuring, in a statistically consistent way, how closely a machine can match human performance in a specified 'intelligent' task.

The imitation game is played by three people: a man (M), a woman (W), and an interrogator (P). P is kept apart from M and W, and the goal of P in the game is to identify who is the man (and who is the woman) by communicating with M and W only in written. P can ask M and W any questions. However, M's task in the game is to cause P to make the wrong identification, while W's task is to help P to make the correct identification. Therefore, M will probably lie most of the time. Suppose we play the imitation game n times, and $c \leq n$ is the number of times that P makes the correct identification at the end of each game. Then, Turing's idea is to focus on the following questions:

'What will happen when a machine takes the part of [M] in this game?' Will the interrogator decide wrongly as often when the game is played like this as he does when the game is played between a man and a woman? These questions replace our original, 'Can machines think?' [306, p. 434]

In other words, our new question concerns whether and how the value of c changes when a machine replaces M within the game. Now that we have in mind what the imitation game amounts to, let us reflect on the precise meaning of the first one of the two quotes. Turing's initial concern is whether it is possible for a machine to think. In particular, this question is a *conceivability* question, asking whether we can imagine conditions under which we would be willing to ascribe thought to a machine. However, in order to answer conceivability questions it is first necessary to agree on the meaning of the words framing such questions. The kind of strategy often adopted by philosophers to answer conceivability questions, known as 'philosophical analysis' (or 'conceptual analysis') in the strict sense [262, 293], is what Turing refers to ironically as a sort of Gallup poll.[1] Now, the idea that philosophical analysis is what we need in order to know whether a machine can think is immediately

---

[1] Although we entirely agree with Alan Turing that philosophical analysis seems to be (sometimes) nothing but an obscure kind of statistical survey concerning linguistic usages, this critical view of conceptual analysis is not a necessary requirement of the argument we are developing. In particular, we are aware of the fact that some philosophers would judge such a view of analysis to be unfair.

discarded by Turing as *absurd*. At this point, it may seem that Turing's rejection of the philosophical analysis approach to the problem of machine intelligence is questionable, since it is not supported by an explicit argument. In particular, it seems difficult to deny that philosophical analysis (or statistical surveys of linguistic usage) is a reasonable way of addressing conceivability questions. However, Turing is very clear in motivating the resort to the imitation game. He does not propose to use that game (rather than a statistical survey) to answer the question about thinking machines, but he proposes instead to *replace* the question about thinking machines by a different question, "which is closely related to it and is expressed in relatively unambiguous words". In other words, what Turing discards as absurd is the idea of approaching the study of machine intelligence by means of conceivability questions (hence philosophical analysis), and he proposes instead to frame that study in terms of *operational* questions, such as those related to the imitation game. While it is not really important to agree on the exact significance and impact of the imitation game, it is instead fundamental, in the economy of Turing's argument, to realize the importance he attaches to an operational approach to questions concerning machine intelligence [243, pp. 15–17].

Now, what does an operational approach to the study of machine intelligence amount to? If we accept Turing's imitation game as a *paradigm* (in the Kuhnian sense [181]) of a research strategy in artificial intelligence, then one important lesson we can draw from it is that AI research, as originating from Turing's pioneering work, is inherently *behavioristic*. That is to say, the notion of intelligence underlying the science of intelligent machines is framed in terms of what we commonly regard as intelligent behavior, not in terms of what we know (or suppose) to implement intelligent behavior in (cognitively evolved) natural organisms.

This behavioristic view of AI has not gone without criticisms, in the last decades, within the philosophy of AI. While an exhaustive treatment of such criticisms goes beyond the scope of the book, it is worth spending just a few words on one influent criticism, the Chinese room argument, developed by John Searle [273]. Searle's original aim is to show, by means of a thought experiment known as 'Chinese room' (i.e. by means of philosophical analysis), that no computational notion can be an adequate notion of intelligence, and (indirectly) that Turing's test cannot be adequate as an intelligence test. However, the Chinese room argument can be read more generally as aimed at showing that no behavioristic notion of intelligence can be adequate. The reason for this is that the argument relies on the idea that a certain system can be behaviorally identical to another system, where the former lacks intelligence and the latter is instead truly intelligent. Therefore, the mere behavior of a certain system is not sufficient (in Searle's view) as evidence that the system is intelligent. Although several objections could be (and have been) raised against various assumptions underlying the Chinese room argument [243, p. 19], the very aim of this argument seems to be seriously misguided. As we saw above, the main idea in Turing's paper is to *replace* questions about machine intelligence by operational questions, framed in behavioristic terms, and not to *answer* the former questions in behavioristic terms. In other words, the behavioristic view of AI does not rely on a behavioristic analysis of the psychological notion of intelligence, but it consists instead in a paradigm shift, switching from a folk-psychological notion of intelligence to a rigorously operational, behavioristic notion. As in genuine paradigm shifts, it does not come to answering old questions in new ways, but it comes instead to establishing new questions as the proper objects of scientific inquiry. Therefore, one major flaw of Searle's argument is to address Turing's test from the point of view of philosophical analysis, which is exactly the perspective that Turing takes care to *eliminate* from the investigation of machine intelligence.

Of course, one may claim that Turing's rejection of philosophical analysis is unjustified, or at least questionable, and that such rejection undermines the philosophical interest or plausibility of Turing's reflection. We will not argue at all against such a move. Instead, we will just dismiss it, appealing to a sort of principle of authority (hence the name of the argument presented in this section). If it comes to choosing between Turing's 'unphilosophical analysis' of machine intelligence, which has been inspiring decades of real scientific research, and Searle's philosophical analysis (of his own intuitions, rather than of real AI results), we definitely opt for the first choice. Significantly, the textbook which is most widely employed today in academic AI courses, whose authors are Stuart Russell and Peter Norvig, adopts a characterization of AI as the investigation of artificial systems that *behave rationally* [269, Section 1.1], where the notion of behavior is explicitly contrasted to thought and other psychologically characterized processes, and rationality is carefully distinguished from its particular implementation in humans. Even more significantly, a philosophical book such as Tim Crane's *The Mechanical Mind* [54] adopts instead a characterization of AI as "the view that computers can think" [54, p. 117], while it interprets Turing's test as assuming that, if a machine is able to play successfully the imitation game, "then we can say that the machine is thinking" [54, p. 117]. But as we should have understood at this point, all we can say if we stick to Turing's argument—hence, if we reason from the perspective of real AI research—is that, if a machine is able to play successfully the imitation game, then a machine is able to *behave* intelligently.

An important objection against the eliminative philosophy of AI we are advocating could be that the very history of AI contradicts our central thesis. In this respect, an illustrative example is the influential work by Herbert Simon (and some of his colleagues) on reasoning and problem-solving, starting from the late Fifties, which was of great importance to the progress not only of AI, but also of cognitive psychology [229]. In fact, Simon believed that AI techniques provide plausible models of human cognition, since the latter consists (in his view) of symbolic computations. However, in order to understand that these remarks do not affect eliminativism, we only need to keep in mind that the eliminative philosophy of AI does not deny the possibility—or even the reality, since it is a historical fact—of a fruitful interaction and cross-fertilization of AI and cognitive psychology (or cognitive neuroscience), but it only denies the methodological necessity of such an interaction. By the way, looking back at the history of AI, Russell and Norvig themselves remark that, while a failure to distinguish between AI techniques and human cognition was typical of the early days of AI, as AI was not yet an established science, in more recent times "this distinction has allowed both AI and cognitive science to develop more rapidly"[269, p. 4].

7.2.2   The Argument from Scientific Practice

The second argument for the eliminativist philosophy of AI, which we call 'the argument from scientific practice', is a more lengthy and complex one. In fact, it is given by the bulk of Chapters 2–6, which are nothing but a piece of contemporary machine learning/AI research. The concepts, methods, and principles presented in those chapters are fairly representative of current mainstream AI, since probabilistic graphical models and related issues in statistical learning are today a thriving research area in the field. As the reader went through the material collected in the previous chapters, it should have been apparent how both theoretical foundations and experimental results in statistical learning and probabilistic graphical models do not depend or rely in any significant way on what we know (or suppose to know) about natural minds. Therefore, the basic strategy behind the argument from scientific practice is just to exhibit a piece of scientific research as evidence for what scientific research really is, notwithstanding any armchair conception of science.

Actually, given the general philosophy of AI outlined in Section 7.2.1, the investigations presented in chapters 2–6 can be viewed as a particular, contemporary application of the paradigm provided by Turing's use of the imitation game, in the following sense. Just as Turing replaces the question 'Can machines think?' by the imitation game, similarly there are a number of philosophical questions that can be fruitfully operationalized as research questions in machine learning. In particular, while Turing's inspiring question is a major issue in the philosophy of mind, we believe that machine learning research provides a natural and powerful way of operationalizing *epistemological* questions. These questions will be addressed in Section 7.4.

7.3   From Cognitive Science to Cognitive Technology

From the arguments proposed thus far, it may seem that the view of AI we are advocating is mainly limitative. On the one hand, that view is somewhat deflationary concerning the relationship between AI and the study of human cognition; on the other hand, the eliminative philosophy of AI strongly discourages any attempt to look at theories and results in AI through the lenses of folk-psychological (or psychological) concepts. However, the positive implications of the perspective adopted in this work lie in clearing the ground for a view of AI as a science investigating and developing *cognitive technologies*. While AI techniques do not necessarily teach anything about natural cognition, they contribute indeed to extend and deepen natural cognition in many significant ways. We now provide two illustrations of this thesis, the first relating to the particular contribution offered in this work, the second relating to some achievements and research directions to be found in the history of AI (and related areas of computer science).

The main problem considered in the book is the problem of learning (large and complex) joint probability distributions from data. The general framework used to deal with this problem is the framework of probabilistic graphical models. This kind of learning task would be tremendously difficult to be accomplished in practice by a human *unaided by a suitable machine*. In particular, it would be prohibitively expensive for a human to perform this task in a manner that is warranted to be consistent with a specified set of statistical axioms and aims, as it is instead the case with the algorithms employed to learn probabilistic graphical models. Given this, the real power and interest of statistical learning methods in AI is that they allow humans, *if aided by suitable machines*, to tackle problems they would not be able to tackle (in a statistically consistent way) without those machines. In other words, statistical learning methods such as the ones considered in the book are properly cognitive technologies (rather than cognitive models), i.e. tools for extending human cognition to novel domains. The view of AI techniques accompanying the results presented in this work is that it does not make sense to ask, about those techniques, 'Are they intelligent?'. The question which is really worth asking about AI is instead 'What can humans accomplish, once AI delivers such and such techniques?'.

The second illustration of the view of AI as cognitive technology derives from two examples of the way that AI can contribute to extend human cognition in different domains. The first example is given by theorem provers and their use in mathematics:

> Theorem provers have come up with novel mathematical results. The SAM [...] program was the first, proving a lemma in lattice theory [...]. The AURA program has also answered open questions in several areas of mathematics [...]. The Boyer-Moore theorem prover [...] was used by Natarajan Shankar to give the first fully rigorous formal proof of Gödel's Incompleteness Theorem [...]. The OTTER program is one of the strongest theorem provers; it has been used to solve several open questions in combinatorial logic. The most famous of these concerns Robbins algebra. In 1933, Herbert Robbins proposed a simple set of axioms that appeared to define Boolean algebra, but no proof of this could be found (despite serious work by several mathematicians including Alfred Tarski himself). On October 10, 1996, after eight days of computation, EQP (a version of OTTER) found a proof [...]. [269, p. 309]

The number and worth of the results achieved by means of theorem-proving programs gives much more substance to the view of AI techniques as tools for extending human cognition, than it gives to the view of those techniques as models of human cognition. In fact, while those results are now an important part of human mathematical knowledge, it is not at all clear whether the way they were achieved entails anything about how the human mind goes about proving theorems. More importantly, the capability of theorem-proving programs to achieve those results does not depend at all, as far as we know, on their being or not being similar to human strategies for proving theorems.

A second example of the influence of AI (and related areas of computer science) on the growth of human knowledge is given by the young, interdisciplinary research field of bioinformatics [49]. In this domain, a number of reasons—such as the high dimensionality of the data to be dealt with, its huge amount, or the fact that these data are generally incomplete and noisy—make it necessary to resort to sophisticated machine learning techniques and statistical methods in order to aid the human attempt to analyze and explain the collected data, and to predict new phenomena based on previously observed ones. Concerning the capability of computational techniques to lead to (otherwise unattainable) discoveries in molecular biology, it is sufficient to note just that, for example, a prominent journal in the field such as *Bioinformatics* devotes a special section of each issue, called 'Discovery Notes', to discoveries achieved by means of computational methods.

7.4 Statistical Machine Learning and the Philosophy of Science

One major question in the philosophy of science is whether there can be a normative theory of scientific discovery, as opposed to a normative theory of justification. In the twentieth century, two representative stands on this subject have been taken for example by Karl Popper [246] on the one hand, who denied the possibility of a logic of scientific discovery, and on the other hand by Herbert Simon [279, 187], who strenuously advocated the plausibility of such a research program, based on preliminary results achieved in artificial intelligence. Although Simon regarded computational models of discovery as plausible models of *human* discovery [229], the significance he attaches to artificial implementations of discovery strategies is not dependent on Simon's views concerning human psychology. While the possibility of a logic of scientific discovery has been the subject of much debate among philosophers, it is highly instructive to consider how the basic insights of a normative theory of discovery have been turned into a thriving research program by computer scientists, namely into the research field which is now known as machine learning.

Probabilistic graphical models are among the most flexible machine learning methods developed in the last decades. While these models allow for an efficient representation of joint probability distributions and automated inference over stochastic domains, one of their main limitations lies in the high computational cost of learning them from data, which makes it infeasible to learn them in domains involving relatively large numbers of variables.

In Section 7.4.1, we wish to stress those features of probabilistic graphical models (and of machine learning formalisms in general) that make them plausible as computational counterparts of scientific 'theories'. To this aim, after noticing how these models allow for the main kinds of inference that have traditionally been regarded as a trademark of scientific theories (namely induction, deduction, and abduction), we focus on the idea of learning statistical models from data. The analysis will show how machine learning is casting new light on traditional problems in the philosophy of science. In particular, Section 7.4.2 explores the implications of some results in statistical learning with respect to the role of simplicity in theory choice, while in Section 7.4.3 we argue that one philosophical lesson we can draw from machine learning research is that scalability (as a formal property of learning methods) can play a fundamental role in making scientific discovery effective.

7.4.1 Machine Learning as a Logic of Scientific Discovery

First of all, let us state what we mean by 'logic of scientific discovery'. Based on Herbert Simon's usage of that phrase, the logic of scientific discovery is meant as *a normative investigation of the inference processes leading to the introduction of (novel) scientific theories.* Concerning this notion, it is very important not to confuse the logic of scientific discovery with the psychology of scientific discovery. While Simon believed that AI techniques provide plausible models of human cognition, the viability of the project of a normative theory of discovery does not depend at all on a psychologistic view of AI.

Of course, one may wonder whether such a normative investigation of scientific discovery is possible at all. A possible strategy for answering this question might be to provide a philosophical argument aimed at showing that the project at issue is indeed theoretically sound (and practically viable). However, this will not be our strategy. Rather than using philosophical arguments in order to advocate the plausibility of a logic of scientific discovery, we simply suggest how machine learning can be interpreted broadly as a project in the logic of scientific discovery. The reason for taking this stance is that, from the philosophical perspective we are assuming, the logic of scientific discovery is not merely a philosophical project, but it is in fact a mature (and thriving) scientific discipline.

Machine learning is the theoretical and experimental study of computational systems whose performance at specific tasks improves with experience [217]. 'Computational systems' means (more or less complex) combinations of algorithms, typically implemented in real computer programs. Performance is usually measured by evaluation metrics that depend on the considered tasks, such as classification accuracy for pattern recognition. Experience is given by a collection of data items. Such data points can take the form of vectors of features (i.e. variables), sequences, graphs, or other suitably formalized objects. In intuitive terms, the aim of a machine learning system is to acquire the capability of solving a certain class of problems by being trained on a set of solved problems (belonging to that class).

In a sense, machine learning algorithms deliver *models* of the data they are trained on. But how should we regard those models, from the point of view of the philosophy of science? Our claim is that such models are nothing but computational counterparts of what we commonly regard as scientific *theories*. 'Theory' is a much debated term in the philosophy of science, and several different views of its meaning have been advocated thus far [28]. Philosophical notions of theory range, for example, from the logical-empiricist conception of a formal system [224], i.e. a set of sentences expressed in first-order logic, to the structuralist idea that a theory should be identified with the set of its models [289], in the strict (model-theoretic) sense of semantic structures [56]. Now, let us reflect on the following passage, drawn from Ian Witten and Eibe Frank's introduction to data mining:

What is learned by a machine learning method is a kind of "theory" of the domain from which the examples are drawn, a theory that is predictive in that it is capable of generating new facts about the domain—in other words, the class of unseen instances. Theory is a rather grandiose term: we are using it here only in the sense of a predictive model. Thus theories might comprise decision trees or sets of rules—they don't have to be any more "theoretical" than that. [317, pp. 179–180]

The remarks just quoted contain a simple yet fruitful insight. In our view, a possible reason why no general consensus has been reached by philosophers in analyzing the notion of theory lies in the fact that the philosophical aim has generally been to explain *what a theory is*, rather than *what a theory is useful for*. Strictly speaking, the latter question does not even have a precise meaning until we answer the former. Nevertheless, while the former question has in fact no universally accepted answer as yet, it would be hard to deny that *inference* (hence explanation and prediction) is the main purpose of scientific theories. In fact, it is clear that any notion of theory should be consistent at least with the following fact: theories are tools for performing (different kinds of) inference. Although a loose notion of theory as an inferential device (or a 'predictive model', as Witten and Frank put it) may fall short of the expectations of philosophers of science, that notion has the important effect of encouraging us to turn our interest from the reflection on the notion of theory to the identification of rational strategies for developing (extending, revising, etc.) scientific theories.

Viewing theories broadly as inferential devices allows us to realize how machine learning methods are nothing but methods for automating the construction of scientific theories, since any machine learning method is aimed at supporting some kind of inference. We think that the success of machine learning (and of AI in general) in opening new perspectives for the application of computational methods to challenging scientific problems—such as theorem-proving and bioinformatics—makes a fairly strong case for the plausibility of viewing machine learning as a logic of scientific discovery.

7.4.2 Simplicity Reconsidered

A common view in the philosophy of science maintains that, other things being equal, simpler theories should be preferred over more complex ones [11, 14]. The puzzle concerning this view is that no account of simplicity has ever been able to reach universal consensus among philosophers. Indeed, several notions of simplicity should be kept distinct from one another, such as *syntactic* simplicity, relating to the form of scientific theories, and *ontological* simplicity, relating to the objects postulated by theories. In this section, we propose to reflect on the syntactic notion of simplicity, based on the theory presented in the previous sections. However, our aim is not to establish a philosophical account of simplicity. Our strategy will be instead to focus on a notion of simplicity grounded in the theory of statistical learning, and to stress the importance of the role played by simplicity in making learning (i.e. induction) effective. While this may not answer many open questions in the philosophical debate about simplicity, it should nevertheless help to understand what part of the philosophical puzzle can be saved as a meaningful problem in the methodology of science. The theory related to the MDL principle (discussed in Section 2.4.1) is of fundamental importance to the problem we are going to address. In particular, the present philosophical reading of that theory is nothing but a way of reformulating it in less technical terms, which does not add anything new to what is regarded as 'received wisdom' in the field of statistical learning. This means that, if the philosophical reader feels that something is wrong with the ideas expressed in this section, the proper way of refuting them would be by refuting the content of Section 2.4.1.

The reason for using the MDL heuristic function is that, when learning a Bayesian network, in order to avoid overfitting the data we need to introduce, in our evaluation function, a careful tradeoff between the likelihood and the complexity of the evaluated models (where complexity is measured by the number of parameters contained in a given model). In other words, such a tradeoff within the scoring function is necessary for the learned model to *generalize well* to new data. Now, since the models being evaluated are hypotheses concerning some specified domain, and since the number of model parameters is nothing but a measure of the syntactic simplicity of those hypotheses, one general implication of the theory of Section 2.4.1 is that an appropriate weighting of the simplicity of hypotheses is necessary in order to find hypotheses that generalize well to future data. That is to say, when formulating hypotheses, *taking into account (syntactic) simplicity plays a precise role in making induction effective.* An interesting question to ask (and which was already answered in Section 2.4.1) is then: why is simplicity so effective? The simple moral to be drawn from derivation 2.22 is that minimizing the description length of a hypothesis h, i.e. maximizing its 'simplicity', means nothing but maximizing the prior probability of h. That is to say, the intimate mathematical connection between simplicity (i.e. description length) and prior probability explains the contribution given by simplicity to the success of induction (i.e. learning): other things being equal, a simpler hypothesis will have a higher posterior probability than a more complex one. The same philosophical point is also made in [92] by discussing Akaike's statistical results [4], which are very similar in their spirit to the meaning of the MDL principle.

7.4.3 Scalability as an Epistemic Virtue

From the perspective of the philosophy of science, one interesting point emerging from the research presented in this book is the following. As we saw, the key advantage of some learning methods over different ones may be given by the stronger *scalability* properties of the former, which makes it easier to learn a statistical model in high-dimensional domains. The philosophical reader should not take this advantage to be a merely practical (and relatively unphilosophical) one. In fact, in many important cases (e.g. in bioinformatics or Web mining) higher scalability means capability of allowing for induction, where less scalable techniques may defuse any attempt to induce a model from data. In other words, improving scalability means extending our attempt of inducing novel scientific theories to domains where induction was not even conceivable before. In this sense, one philosophical lesson we can draw from the effort of designing machine learning models that are suitable for application to high-dimensional tasks is that scalability should be regarded as an *epistemic virtue* in its own right, i.e. as one of the basic parameters we ought to consider when evaluating scientific theories, on a par with other virtues such as simplicity or consistency (and, of course, besides explanatory and predictive power) [247, 248].

7.5 Final Remarks

This chapter was meant to offer a philosophical reflection on some issues concerning the relationships between statistical machine learning (and, in particular, research on probabilistic graphical models) and the cognitive sciences. First, two arguments— i.e. the argument 'from authority' and the argument 'from scientific practice'—have been put forward in order to locate machine learning—and AI in general—within the framework of the cognitive sciences in a sensible way. We believe that the two arguments make a fairly strong case against all those who maintain that the ultimate success of AI research depends on whether natural cognition can be successfully explained in terms of computational processes. On the other hand, the *pars construens* of our proposal for the philosophy of AI consists in shifting from a cognitive science perspective to a cognitive technology perspective, i.e. in viewing machine learning and AI as normative—rather than descriptive—sciences. In this respect, we argued that machine learning is establishing itself as a full-fledged implementation of the philosophical project that Herbert Simon called 'logic of scientific discovery'. In particular, such a realization of Simon's project is not only consistent and theoretically sound, as shown by the theory of statistical learning, but it is also practically effective, as the technological impact of machine learning research (and AI in general) is no longer a mere promise, but it is an extremely important part of the world we are living in today. A paradigmatic application of this perspective to a couple of philosophical problems—involving the role of simplicity in theory choice and the traditional class of epistemic virtues—has been finally offered, by explicitly deploying the statistical machine learning framework underlying this book. In particular, we showed how an elegant and mathematically robust solution for the philosophical puzzle of simplicity can be drawn from statistical learning theory, in particular from an analysis of the MDL principle and its role in learning algorithms for probabilistic graphical models. Also, we claimed that scalability, as a formal property of induction techniques for scientific theories, should be recognized to play a very important role in making induction effective, and hence in allowing science to develop more rapidly.

Chapter 8
Conclusions

8.1 Hybrid Random Fields: Where Are We Now?

The time has come to summarize the main contributions we have attempted to make throughout the book. To this aim, let us look back at each one of the main chapters, in turn:

- Chapters 2–3 presented the main theoretical concepts and engineering techniques related to the representation and estimation of joint probability distributions in Bayesian networks and Markov random fields. After reading them, the reader should have acquired some of the basic competences needed on the one hand to implement and use directed and undirected graphical models in application domains involving discrete variables only, and on the other hand to read more advanced literature dealing with learning techniques for (discrete) probabilistic graphical models;

- Chapter 4 introduced the hybrid random field model, exploring its mathematical properties and explaining how it can be used for representing joint probabilities (via Gibbs sampling) and pseudo-likelihood functions (via simple factorization into local conditional distributions), for estimating pseudo-likelihood distributions in large-scale domains involving discrete variables (via suitable parameter and structure learning algorithms), and for accomplishing arbitrarily shaped inference tasks (again, via Gibbs sampling). After reading this chapter, readers should be able to exploit hybrid random fields as an alternative for virtually any learning and reasoning application addressed through the techniques drawn from Chapters 2–3;

- Chapter 5 explained how directed, undirected, and (more explicitly) hybrid graphical models can be applied to domains involving continuous-valued variables. To this aim, three different families of (conditional) density estimation techniques were presented, i.e. parametric techniques (based on Gaussian models), semiparametric techniques (based on the nonparanormal model), and nonparametric techniques (based on kernel methods). On the other hand, the issue of structure learning in hybrid random fields was tackled from the perspective of continuous application domains, which required to suitably adapt the Markov

A. Freno and E. Trentin: Hybrid Random Fields, ISRL 15, pp. 163–167.
springerlink.com   © Springer-Verlag Berlin Heidelberg 2011

Blanket Merging algorithm developed in Chapter 4. This chapter should have provided the reader with some tools needed for extending probabilistic graphical modeling to various kinds of densities underlying high-dimensional, continuous random vectors;

- Chapter 6 presented a number of applications of probabilistic graphical models to feature selection, pattern classification, and link prediction tasks, featuring both synthetic and real-world benchmarks involving discrete and continuous variables. The considered applications allowed to evaluate the behavior of hybrid random fields from the points of view of prediction accuracy and learning scalability as well, with very promising results. After considering this chapter, readers should have gained insight into a few representative kinds of applications allowing for learning and inference through probabilistic graphical models, and especially into the practical advantages of hybrid random fields as an alternative to more traditional graphical models;

- Finally, Chapter 7 investigated the relationships existing between the theoretical and application-oriented framework inspiring the research presented in this book and some epistemological questions traditionally dealt with in the philosophy of cognitive science. After delving into the arguments put forward in this chapter, the philosophically-minded reader should have realized the relevance of even the most overly technical sections of the book for a number of interesting issues in epistemology and cognitive science.

8.2   Future Research: Where Do We Go from Here?

Now that the big picture has finally emerged concerning the theoretical results and practical tools presented in this book, let us sketch just a couple of directions for further research in probabilistic graphical models, with special emphasis on the framework provided by hybrid random fields. Section 8.2.1 outlines some open problems in statistical relational learning, whereas some final thoughts on future perspectives for work in (nonparametric) density estimation are offered in Section 8.2.2.

8.2.1   Statistical Relational Learning: Some Open Questions

Statistical relational learning (SRL) is a young research field in machine learning and knowledge representation, which is motivated by the attempt to bring statistical learning methods to bear on problems that involve not only uncertainty, but also relational structure [108, 68, 60, 109]. The motivation for the attempt arises from observing that the structure of many real-world problems provides information that is both uncertain and relational. Notable examples are given by social network analysis, web mining, and image understanding. In recent years, an approach to SRL that has attracted important research efforts consists in complementing (graph-based) statistical techniques with formalisms derived from (or interpretable in terms of) first-order logic (FOL), or fragments thereof [32]. Such a research direction has lead to a number of different models, like the ones proposed for example in

[99, 166, 107, 222]. Unfortunately, at least one difficulty associated with the usage of these models has prevented thus far their wide adoption in real-world applications, namely the higher formal complexity of the employed mathematical tools, as compared to the formalisms employed in standard probabilistic graphical models. More importantly, the computational complexity of estimating the structure (i.e. the graph) of these models from data is one of their major limitations. For example, learning the structure of probabilistic relational models [99] is at least as hard as learning (the structure of) standard Bayesian networks, which is known to be NP-hard.

An interesting attempt in SRL research is made in [256], where the *Markov logic network* (MLN) model is proposed, aimed at satisfying two desiderata. On the one hand, the theory of MLNs is meant to subsume both FOL and statistical learning, hence preserving the aims and scope of previous SRL models. On the other hand, MLNs explicitly attempt to be simpler to understand and apply than other competing models of the same family. The basic idea behind MLNs is to start from a knowledge-base expressed in FOL, and then to learn a set of weights for the clauses in the knowledge-base. Such weights are learned by optimizing the parameters of a Markov random field such that the feature functions in the MRF correspond to groundings of the clauses in the knowledge base. Given the learned weights, probabilistic inference can then be performed in the modeled domain with respect to queries expressed in first-order logic. In MLNs, FOL provides the user interface, while MRFs are used to implement statistical learning and inference. Since an initial (non-weighted) knowledge-base is needed in order to learn and apply the model, this feature of MLNs is responsible both for one of their major selling-points, and for one of their major limitations. When domain knowledge is available in logical form, MLNs are very easy to deploy, and they offer an elegant way to refine that knowledge in a statistically sound way. But when we have no logical prior knowledge concerning the application domain, constructing a knowledge-base (even an incomplete one) can be very challenging, and the task inherits all the difficulties of traditional inductive logic programming (ILP). When no prior knowledge is available, a typical way of addressing the construction of the knowledge-base in MLNs, pursued for example in [256], is by employing an off-the-shelf ILP technique, such as the CLAUDIEN system [59]. The ILP approach raises two problems. First, learning in ILP systems can be very slow when dealing with high-dimensional domains, due to its computational cost. Second, such systems can be pretty difficult to deploy for the user, due to a large number of parameters that must be tuned manually by the user. More dedicated algorithms for learning the structure of MLNs, either from scratch or from an initial knowledge-base, are developed in [172, 216, 173]. Although such algorithms lead to improvements with respect to systems like CLAUDIEN or FOIL [250], the computational burden imposed for learning a model on domains of reasonable dimensionality can be extremely high.

In sum, two important challenges that have not been met yet by logic-based approaches to SRL concern first the computational efficiency of the learning methods, and second the capability of such methods to work well even in the limit case where no prior knowledge is available (or at least easily representable in logical form) concerning the application domain. On the other hand, a mathematical issue that deserves serious investigation concerns the implications of some restrictive assumptions that SRL models usually make with respect to the used fragment of FOL. For example, MLNs (at least in their standard usage [256]) assume that the quantification domain of the logical variables has finite cardinality. While some research has tried to address the possibility of relaxing the finite domain assumption [150, 242, 280], *finite* model theory [75, 118] (i.e. the restriction of model theory [56] to finite structures) has established itself as an independent and burgeoning research field in mathematical logic. We believe that exploring the implications of finite model theory research for SRL methods making the finite domain assumption might provide fruitful insights for future developments in SRL.

Besides the issue of learning scalability, an interesting question to ask concerning any given SRL framework is the capability of the proposed framework to deal effectively with *continuous* variables (i.e. with quantities ranging over uncountable sets of values). This issue is an open challenge in SRL, since virtually any SRL model inherits the intrinsic difficulties associated with the usage of classic (probabilistic) graphical models. While the issue of scalability has gained increasing attention in recent years [228], the problem of dealing with continuous variables efficiently while avoiding to make restrictive (and sometimes unjustified) assumptions on the nature of the modeled phenomena (such as Gaussianity of the underlying distribution) has been tackled only sporadically in the graphical models community [137, 197, 97]. This state of affairs in probabilistic graphical models research is astonishing if we look at it from the perspective of established statistical machine learning and pattern recognition [73], where continuous quantities have traditionally been the focus of most research efforts. Therefore, we believe that progress in this direction would offer to the developed framework a dramatic advantage over existing SRL approaches, which are mainly limited to discrete application domains.

Based on the analysis sketched above, two desiderata that a novel SRL approach should be expected to satisfy are on the one hand the scalability of learning and reasoning, and on the other hand the capability of dealing with continuous domains. As described throughout this book, the presented methods for learning hybrid random fields are able to outperform standard learning techniques for Bayesian networks and Markov random fields in terms of scalability properties, while preserving (and often even improving) the predictive accuracy of the alternative models. Moreover, all benchmarks considered thus far for hybrid random fields did not offer any prior knowledge for aiding the model selection process. Since HRFs have not been extended thus far to SRL tasks, this kind of graphical models seems to be a fairly promising starting point for further research in SRL. The reason is that, while in the non-relational setting HRFs satisfy both the requirements of learning scalability and of robustness to missing prior knowledge, in the relational setting they offer a novel foundation (just as Markov random fields offered a foundation for Markov logic) for building an alternative SRL paradigm which may be more promising than the existing ones.

Concerning instead the treatment of continuous variables, Chapter 5 shows how hybrid random fields (and graphical models in general) can be used to model (continuous) density functions while avoiding parametric assumptions. Scalability is an even more challenging requirement for continuous (rather than for discrete) graphical models, and hence the perspectives for improvement are extremely open on this front. Given the potential advantages that effective learning techniques for continuous graphical models would offer (besides discrete models) as a basis for relational real-world applications, we believe that the scalable framework for continuous pseudo-likelihood estimation presented in Chapter 5 provides an interesting starting point for research in SRL over continuous domains.

8.2.2 Nonparametric Density Estimation: Beyond Kernel Machines

Another issue that seems to be pretty open to further research developments is the general problem of nonparametric density estimation. While the main nonparametric approach explored thus far in the literature exploits kernel-based methods, as reviewed in Section 5.5, addressing this task by means of different mathematical tools would be an intriguing research challenge. In fact, one major limitation of kernel-based estimators is that they are relatively inefficient at prediction time, due to their memory-based nature. And, as we saw throughout this book, scalability is to be regarded not just as a nice practical advantage of some estimation methods over alternative ones, but rather as a crucial requirement of any learning and inference technique aiming at real-world applicability.

In recent years, one nonparametric approach trying to overcome the limitations of standard kernel-based methods has been deploying artificial neural networks, where learning relies either on empirical estimates of cumulative distribution functions [203], or on an 'unbiased' training supervision offered by suitably exploited Parzen window and k_n-nearest neighbor estimators [297, 300]. Both strategies—i.e. the edf-based technique and the unbiased labeling-based one—have been tested only on univariate benchmarks. While the latter technique may reveal to be effective even for multivariate distributions, it is not at all clear whether and how the approach based on empirical distribution functions may be extended to multivariate data. On the other hand, the methods based on the unbiased labeling drawn from Parzen window or k_n-nearest neighbor estimators are relatively slow during the training phase, and their computational advantages only show up at test time. While this is already an important improvement over traditional kernel-based density estimators, it may not be enough if we look at the constraints imposed by several large-scale, real-world applications. Given the crucial role played by (conditional) density estimation within hybrid random fields (and probabilistic graphical models in general), we believe that progress along this direction could make a fundamental contribution to the research presented in this book.

Appendix A
Probability Theory

A.1 Random Variables

One fundamental notion in probability theory is the notion of a *sample space* [233, 8]. Given a 'random experiment' E, the sample space Ω associated with E is the set of all possible outcomes $e_1, \ldots, e_n$ of the experiment. Intuitively, one may think of a random experiment as the operation of picking out an element from a set and determining the value of a specified attribute for that element. For example, we may pick a person from a population and check whether that person is European or not; or, we may measure the weight of that person, where the value of the weight (as expressed e.g. in kilograms) lies in the interval $(0, \infty)$. In the first case, we are dealing with a discrete sample space $\Omega = \{e_1, e_2\}$, such that e_1 corresponds to being European and e_2 corresponds to not being European; the sample space for the second experiment is instead a continuous set $\Omega = \{x : 0 < x < \infty\}$. An experiment in the considered sense is not well-defined until its sample space has been identified. Given a sample space Ω, an *event* is any set E such that $E \subseteq \Omega$. An event E is said to be *elementary* if $|E| = 1$. Moreover, two events E_1 and E_2 are said to be *mutually exclusive* (or, simply, *disjoint*) if $E_1 \cap E_2 = \emptyset$.

Another basic concept is the concept of random variable:

Definition A.1. Given a sample space Ω, a *random variable X* is a function $X : \Omega \rightarrow \mathcal{X}$, where $\mathcal{X}$ is called the *space* of X. If $\mathcal{X}$ is countable (i.e. either finite or enumerable [142]), then we say that X is *discrete*, otherwise (i.e. if $\mathcal{X}$ is uncountable) we refer to X as a *continuous* random variable.

In cases where no ambiguity may arise, the space of a random variable is sometimes referred to (improperly) as the *domain* of that variable. When working with a random variable X, we use the notation $X = x$ to refer to the set of all events in the sample space such that X associates value x to anyone of those events. That is to say, $X = x$ denotes the set $\{e_i : X(e_i) = x\}$. Notice that random variables are denoted by capital letters, while values of those variables are denoted by small letters. The phrase 'random vector' refers to a vector of random variables. Hence, if $X_1, \ldots, X_n$ are random variables, the object $\mathbf{X} = (X_1, \ldots, X_n)$ is a random vector. It is common to refer to any given value $\mathbf{x}$ of a random vector $\mathbf{X}$ as a *state* of the random vector.

A.2   Discrete Probability Distributions

We now define the notion of probability function:

Definition A.2. Consider a sample space $\Omega = \{e_1, \ldots, e_n\}$. If $\mathscr{P}(x)$ denotes the powerset of x, i.e. the set $\{y : y \subseteq x\}$, then a function $P : \mathscr{P}(\Omega) \to \mathbb{R}$ is a *probability function* if it satisfies the following conditions:

1. For $1 \leq i \leq n, 0 \leq P(\{e_i\}) \leq 1$;
2. $\sum_{i=1}^{n} P(\{e_i\}) = 1$;
3. For any event E, $P(E) = \sum_{e_i \in E} P(\{e_i\})$.

The ordered pair (Ω, P) is called a *probability space*. For brevity, it is common to say that P is a probability function on the sample space Ω, rather than saying (more properly) that P is a probability function on $\mathscr{P}(\Omega)$. Given Definition A.2, we can easily prove an important theorem:

Theorem A.1. *If (Ω, P) is a probability space (where $\Omega = \{e_1, \ldots, e_n\}$), then (Ω, P) satisfies the following conditions:*

1. *$P(\Omega) = 1$;*
2. *For any E such that $E \subseteq \Omega, 0 \leq P(E) \leq 1$;*
3. *For any pair of mutually exclusive events E_1 and E_2, $P(E_1 \cup E_2) = P(E_1) + P(E_2)$.*

Proof. Let us prove the three conditions stated above in their order:

1. By the third condition in Definition A.2, $P(\Omega) = \sum_{i=1}^{n} P(\{e_i\})$. Since the second condition in Definition A.2 states that $\sum_{i=1}^{n} P(\{e_i\}) = 1$, it follows that $P(\Omega) = 1$;
2. We prove that (a) $P(E) \geq 0$ and (b) $P(E) \leq 1$, respectively:

   a. Since $P(E) = \sum_{e_i \in E} P(\{e_i\})$ and $P(\{e_i\}) \geq 0$ for any i such that $e_i \in E$, we have that $P(E) \geq 0$;
   b. Since $E \subseteq \Omega$, we have that, by the third condition in Definition A.2, $P(E) \leq P(\Omega)$, where $P(\Omega) = 1$. Therefore, $P(E) \leq 1$.

3. First, we have that $P(E_1 \cup E_2) = \sum_{e_i \in E_1 \cup E_2} P(\{e_i\})$. Since E_1 and E_2 are mutually exclusive, it follows that

$$\sum_{e_i \in E_1 \cup E_2} P(\{e_i\}) = \sum_{e_i \in E_1} P(\{e_i\}) + \sum_{e_i \in E_2} P(\{e_i\}) \tag{A.1}$$

   where $\sum_{e_i \in E_1} P(\{e_i\}) = P(E_1)$ and $\sum_{e_i \in E_2} P(\{e_i\}) = P(E_2)$. Therefore, $P(E_1 \cup E_2) = P(E_1) + P(E_2)$.

$\square$

The classical axiomatization of probability theory is due to Andrey Kolmogorov [177], who used the conditions detailed in Theorem A.1 as axioms of the theory.

Given a random variable X, the notation $P(X = x)$ refers to the quantity $P(\{e_i : X(e_i) = x\})$. In particular, we say that $P(X = x)$ denotes the probability that random variable X assumes value x. While the notation $P(x)$ is just a shorthand for $P(X = x)$, the set $\{(x,P(x)) : x \in \mathcal{X}\}$ (where $\mathcal{X}$ is the space of X) is denoted by $P(X)$, and it is referred to as the *probability distribution* (or simply *distribution*) of X.

When dealing with a set of random variables $X_1,\dots,X_n$ (defined on the same sample space Ω), the notation $X_1 = x_1,\dots,X_n = x_n$ (or $X_1 = x_1 \wedge \dots \wedge X_n = x_n$) refers to the set $\{e_i : X_1(e_i) = x_1\} \cap \dots \cap \{e_i : X_n(e_i) = x_n\}$. Thus, $P(X_1 = x_1,\dots,X_n = x_n)$, or $P(X_1 = x_1 \wedge \dots \wedge X_n = x_n)$, denotes the quantity $P(\{e_i : X_1(e_i) = x_1\} \cap \dots \cap \{e_i : X_n(e_i) = x_n\})$. For brevity, $P(X_1 = x_1,\dots,X_n = x_n)$ is often written as $P(x_1,\dots,x_n)$, and it represents the probability that random variables $X_1,\dots,X_n$ assume the values $x_1,\dots,x_n$ respectively. Let us use $\mathbf{X}$ to denote the random vector $(X_1,\dots,X_n)$, and $\mathbf{x}$ to denote $(x_1,\dots,x_n)$. As for the case of a single random variable, we use $P(\mathbf{X})$ to denote the set $\{(\mathbf{x},P(\mathbf{x})) : \mathbf{x} \in \mathcal{X}_1 \times \dots \times \mathcal{X}_n\}$, where $\mathcal{X}_i$ denotes the space of random variable X_i (for $1 \le i \le n$). The set $P(\mathbf{X})$ is referred to as the *joint probability distribution* of $\mathbf{X}$ (sometimes abbreviated as *joint distribution*, or just *distribution*).

We now define the concept of conditional probability:

Definition A.3. Given two random variables X and Y, the *conditional probability* that X assumes value x given that Y has value y, denoted by $P(X = x \mid Y = y)$ or by $P(x \mid y)$, is defined as

$$P(x \mid y) = \frac{P(x,y)}{P(y)} \tag{A.2}$$

provided that $P(y) \ne 0$.

Clearly, Equation A.2 implies that

$$P(x,y) = P(x \mid y)P(y) \tag{A.3}$$

which holds even if $P(y) = 0$. A very useful rule for calculating joint probabilites by means of conditional probabilities is the *chain rule*, which is stated by the following theorem:

Theorem A.2. *Given a set of random variables $X_0,\dots,X_n$, for any combination of values $x_0,\dots,x_n$ of the given variables, the following equality holds:*

$$P(x_0,\dots,x_n) = P(x_0) \prod_{i=1}^{n} P(x_i \mid x_0,\dots,x_{i-1}) \tag{A.4}$$

Proof. The validity of Equation A.4 can be proved by induction on the number of variables. First, consider that, if $n = 1$, then the chain rule reduces to an instance of Equation A.3, i.e. to $P(x_0,x_1) = P(x_1 \mid x_0)P(x_0)$, which we know to be true. Second, suppose that the chain rule is valid for a particular value of n, i.e. that $P(x_0,\dots,x_n) = P(x_0) \prod_{i=1}^{n} P(x_i \mid x_0,\dots,x_{i-1})$, and let us show that the rule also holds for $n+1$. Clearly, $P(x_0,\dots,x_{n+1}) = P(x_{n+1} \mid x_0,\dots,x_n) P(x_0,\dots,x_n)$. This fact, together with the induction hypothesis, implies that $P(x_0,\dots,x_{n+1}) = P(x_0) \prod_{i=1}^{n+1} P(x_i \mid x_0,\dots,x_{i-1})$. Therefore, Equation A.4 is valid for any value of n. $\qquad\square$

An important concept (which is tightly related to the notion of conditional probability) is the concept of statistical independence:

Definition A.4. Two random variables X and Y are said to be *statistically independent* (or simply *independent*) if, for any value x of X and any value y of Y,

$$P(x,y) = P(x)P(y) \tag{A.5}$$

The relationship between statistical independence and conditional probability is expressed by the fact that, if two variables X and Y are independent, then

$$P(x \mid y) = P(x) \tag{A.6}$$

which immediately follows from Definitions A.3, A.4.

Using Equations A.2–A.3, together with the (trivial) fact that $P(x,y) = P(y,x)$, we can prove a useful equality, which is traditionally referred to as *Bayes' theorem* [17]:

$$P(x \mid y) = \frac{P(x,y)}{P(y)} = \frac{P(y,x)}{P(y)} = \frac{P(y \mid x)P(x)}{P(y)} \tag{A.7}$$

Another important rule for computing probabilities is given by the *law of total probability*, which is stated by the following theorem:

Theorem A.3. *Suppose that the sample space Ω is partitioned into n (mutually exclusive and disjoint) events $E_1, \ldots, E_n$. Then, for any event E such that $E \subseteq \Omega$,*

$$P(E) = \sum_{i=1}^{n} P(E \cap E_i) \tag{A.8}$$

Proof. We first note that the following equality holds, based on elementary set-theoretic principles:

$$E = E \cap \Omega = E \cap \bigcup_{i=1}^{n} E_i = \bigcup_{i=1}^{n} (E \cap E_i) \tag{A.9}$$

Now, the third condition in Theorem A.1 implies that

$$P\left(\bigcup_{i=1}^{n} (E \cap E_i)\right) = \sum_{i=1}^{n} P(E \cap E_i) \tag{A.10}$$

Therefore, $P(E) = \sum_{i=1}^{n} P(E \cap E_i)$. $\qquad\square$

An interesting implication of the law of total probability is that the distribution of a random variable can be derived by summing over all values of the remaining variables. Suppose we are dealing with the random variables $X, Y_1, \ldots, Y_n$. Moreover, if $\mathbf{Y} = (Y_1, \ldots, Y_n)$, we use $\mathcal{Y}$ to denote the domain of the random vector $\mathbf{Y}$. Then, for any value x of X, the law of total probability implies the following equality:

$$P(x) = \sum_{y \in \mathcal{Y}} P(x, \mathbf{y}) \tag{A.11}$$

When computed through the relevant values of the joint distribution $P(X, Y)$, the distribution $P(X)$ is usually referred to as *marginal (probability) distribution* of X. Clearly, the marginal probability of x can also be computed as

$$P(x) = \sum_{y \in \mathcal{Y}} P(x \mid \mathbf{y}) P(\mathbf{y}) \tag{A.12}$$

A.3 Continuous Probability Distributions

We first define the concept of *probability density function* (pdf) [233, 82]:

Definition A.5. Given a continuous random variable $X : \Omega \to \mathcal{X}$, a function p defined on $\mathcal{X}$ is a *probability density function* if it satisfies the following conditions:

1. For any x such that $x \in \mathcal{X}$, $p(x) \geq 0$;
2. $\int_{\mathcal{X}} p(x) \, dx = 1$;
3. For any closed set $\mathcal{S}$ such that $\mathcal{S} \subseteq \mathcal{X}$, $P(x \in \mathcal{S}) = \int_{\mathcal{S}} p(x) \, dx$, where $P(x \in \mathcal{S})$ is used as shorthand for $P(\{e : X(e) \in \mathcal{S}\})$.

A pdf defined on a single random variable is called a *univariate* density function, whereas a pdf is said to be *multivariate* when it involves random vectors. As for univariate density functions, also multivariate densities are such that, if $\mathbf{X}$ is a random vector with space $\mathcal{X}$, then:

1. For any $\mathbf{x}$ such that $\mathbf{x} \in \mathcal{X}$, $p(\mathbf{x}) \geq 0$;
2. $\int_{\mathcal{X}} p(\mathbf{x}) \, d\mathbf{x} = 1$;
3. For any closed set $\mathcal{S}$ such that $\mathcal{S} \subseteq \mathcal{X}$, $P(\mathbf{x} \in \mathcal{S}) = \int_{\mathcal{S}} p(\mathbf{x}) \, d\mathbf{x}$.

We also define the notion of *cumulative distribution function*, or *cdf* [233, 82]:

Definition A.6. Given a continuous random variable $X : \Omega \to \mathcal{X}$, the *cumulative distribution function* of X is the function F defined as $F(x) = P(X \leq x)$, where $P(X \leq x)$ is a shorthand for $P(\{e : X(e) \leq x\})$.

The cumulative distribution function is monotonically increasing and right-continuous. Moreover, we have that

$$\lim_{x \to -\infty} F(x) = F(-\infty) = 0 \tag{A.13}$$

and that

$$\lim_{x \to +\infty} F(x) = F(\infty) = 1 \tag{A.14}$$

The mathematical relationship between pdf and cdf is characterized by the following property:

$$F(x) = \int_{-\infty}^{x} p(t) \, dt \tag{A.15}$$

Conversely, we have that

$$p(x) = \frac{\mathrm{d}}{\mathrm{d}x} F(x) \tag{A.16}$$

An important notion—which is tightly related to the cumulative distribution function—is the notion of *empirical (cumulative) distribution function* (edf), defined as

$$F^E(x) = \frac{1}{n} \sum_{i=1}^{n} \Theta(x - x_i) \tag{A.17}$$

where n is the number of training points and Θ is the Heaviside step function, such that $\Theta(x) = 0$ if $x < 0$ and $\Theta(x) = 1$ otherwise. As defined in Equation A.17, the function F^E is discontinuous. If desired, a differentiable, parameterized approximation $F^*(x; \sigma)$ of $F^E(x)$ can be easily formulated using the sigmoid function:

$$F^*(x; \sigma) = \frac{1}{n} \sum_{i=1}^{n} \frac{1}{1 + \exp\left(-\frac{x - x_i}{\sigma}\right)} \tag{A.18}$$

where σ determines the smoothness of the approximating function. The approximation is justified by the following fact:

$$\Theta(x) = \lim_{\sigma \to 0} \frac{1}{1 + \exp\left(-\frac{x}{\sigma}\right)} \tag{A.19}$$

Equation A.17 is commonly known as the *Kaplan-Meier estimate* of the empirical distribution function [160]. A slightly modified definition of $F^E(x)$ is the *mean-rank estimate* [82], given by

$$F^E(x) = \frac{1}{n+1} \sum_{i=1}^{n} \Theta(x - x_i) \tag{A.20}$$

For continuous random variables, the notions of conditional pdf and statistical independence are defined in the same way as for discrete variables, by simply substituting the density function p for the probability function P in Equations A.2, A.5 [86, 87]. In other words, we have that, for any pair of continuous variables X and Y,

$$p(X = x \mid Y = y) = \frac{p(X = x, Y = y)}{p(Y = y)} \tag{A.21}$$

provided that $p(Y = y) > 0$, and that X and Y are said to be independent if, for any values x and y,

$$p(X = x, Y = y) = p(X = x) p(Y = y) \tag{A.22}$$

Similarly, the *marginal density* of x given the value of a random vector $\mathbf{Y}$ is given by

$$p(x) = \int_{\mathcal{y}} p(x, \mathbf{y}) \, \mathrm{d}\mathbf{y} = \int_{\mathcal{y}} p(x \mid \mathbf{y}) p(\mathbf{y}) \, \mathrm{d}\mathbf{y} \tag{A.23}$$

where $\mathcal{y}$ is the space of $\mathbf{Y}$.

One very important notion related to density functions is the notion of *expected value* (or *expectation*) of a function f, denoted by $E[f]$:

Definition A.7. Given a continuous random variable $X : \Omega \to \mathfrak{X}$ (distributed according to $p(X)$) and a function $f : \mathfrak{X} \to \mathbb{R}$, the *expected value* of f is the function

$$E[f] = \int_{\mathfrak{X}} f(x) \, p(x) \, dx \tag{A.24}$$

If we deal with a finite sample $\mathbf{D} = \{x_1, \ldots, x_n\}$, where each instance x_i of X (for $1 \le i \le n$) is drawn from the density $p(X)$, then the expected value of f can be estimated as

$$\widehat{E}[f] = \frac{1}{n} \sum_{i=1}^{n} f(x_i) \tag{A.25}$$

The estimate specified in Equation A.25 converges to $E[f]$ in the limit $n \to \infty$. If we have a random vector $\mathbf{X} = (X_1, \ldots, X_d)$ (with d-dimensional space $\mathfrak{X}$), together with a d-variate function f and a d-variate density p, then $E[f]$ is the multiple integral $\int_{\mathfrak{X}} f(\mathbf{x}) \, p(\mathbf{x}) \, d\mathbf{x}$. The definition of expected value easily extends to the case of a conditional density $p(X \mid y)$, yielding the *conditional expectation* $E[f \mid y]$:

$$E[f \mid y] = \int_{\mathfrak{X}} f(x) \, p(x \mid y) \, dx \tag{A.26}$$

Another central quantity, defined in terms of expected values, is the *variance* of f, denoted by $\sigma[f]$:

$$\sigma^2[f] = E\left[(f - E[f])^2\right] \tag{A.27}$$

where $(f - E[f])^2$ denotes the function $(f(x) - E[f])^2$. As the very name suggests, the variance of a function f indicates how much variability there is in the value of that function around $E[f]$. Of special interest is the variance $\sigma^2[X]$ of a random variable itself, i.e. the value assumed by $\sigma^2[f]$ when we take $f = X$:

$$\sigma^2[X] = E\left[(X - E[X])^2\right] \tag{A.28}$$

where $(X - E[X])^2$ refers to the function $(x - E[X])^2$. The quantity $E[X]$ is usually referred to as the *mean* of random variable X, and denoted by $\mu[X]$. We also define the *covariance* $\sigma^2[X, Y]$ of two random variables X and Y:

$$\begin{aligned}
\sigma^2[X, Y] &= E\left[(X - E[X])(Y - E[Y])\right] \\
&= E[XY] - (E[X]E[Y])
\end{aligned} \tag{A.29}$$

where the values of $X - E[X]$, $Y - E[Y]$, and XY are given by $x - E[X]$, $y - E[Y]$, and xy respectively. Intuitively, the covariance of X and Y measures the extent to which the two variables vary together, i.e. their degree of correlation. Consequently, the covariance vanishes when the variables are independent. If we deal with two random vectors $\mathbf{X} = (X_1, \ldots, X_d)$ and $\mathbf{Y} = (Y_1, \ldots, Y_d)$, then the *covariance matrix* $\Sigma[\mathbf{X}, \mathbf{Y}]$ is a $d \times d$ matrix Σ, such that each entry in Σ is given by $\Sigma_{ij} = \sigma^2[X_i, Y_i]$ (where $1 \le i, j \le d$).

Appendix B
Graph Theory

This appendix reviews some basic notions and a couple of algorithms developed in graph theory. Section B.1 collects the definitions of the graph-theoretical concepts used throughout the book, whereas algorithms for sorting nodes in directed acyclic graphs (while verifying whether the graph is actually acyclic) and finding maximal cliques in undirected graphs are presented in Section B.2.

B.1 Definitions

First of all, we define the notion of *graph*. Graphs can be either *undirected* or *directed*:

Definition B.1. An *undirected graph* $\mathcal{G}$ is an (ordered) pair $(\mathcal{V}, \mathcal{E})$ such that $\mathcal{V}$ is a finite set of so-called *nodes* (or *vertices*) and $\mathcal{E}$ is a set of unordered pairs of nodes. That is, if $e \in \mathcal{E}$, then $e = \{X_i, X_j\}$ and $\{X_i, X_j\} \subseteq \mathcal{V}$. The elements of $\mathcal{E}$ are called *arcs* (or *edges*).

Definition B.2. A *directed graph* $\mathcal{G}$ is an (ordered) pair $(\mathcal{V}, \mathcal{E})$ such that $\mathcal{V}$ is a finite set of nodes and $\mathcal{E}$ is a set of ordered pairs of nodes. That is, if $e \in \mathcal{E}$, then $e = (X_i, X_j)$ and $\{X_i, X_j\} \subseteq \mathcal{V}$.

Both for directed and undirected graphs, we can define the concept of *adjacency* (with respect to a pair of nodes):

Definition B.3. If a graph $\mathcal{G} = (\mathcal{V}, \mathcal{E})$ is undirected, then two nodes X_i and X_j in $\mathcal{V}$ are said to be *adjacent* if $\{X_i, X_j\} \in \mathcal{E}$. If $\mathcal{G}$ is instead directed, then X_i and X_j are said to be adjacent if $\{(X_i, X_j), (X_j, X_i)\} \cap \mathcal{E} \neq \emptyset$.

In undirected graphs, of particular importance are the sets of nodes responding to the definition of *clique*:

Definition B.4. Given an undirected graph $\mathcal{G} = (\mathcal{V}, \mathcal{E})$, a *clique* $\mathcal{C}$ in $\mathcal{G}$ is any subset of $\mathcal{V}$ such that, for any couple of elements X_i and X_j in $\mathcal{C}$ (where $i \neq j$), $\{X_i, X_j\} \in \mathcal{E}$. Furthermore, the clique $\mathcal{C}$ is said to be *maximal* if it satisfies the following condition: $\forall i \, X_i \in \mathcal{V} \setminus \mathcal{C} \Rightarrow (\exists j \, X_j \in \mathcal{C} \wedge \{X_i, X_j\} \notin \mathcal{E})$.

On the other hand, useful notions for characterizing the structure of directed graphs are the notions of *parent*, *child*, and *root*:

Definition B.5. If $\mathcal{G} = (\mathcal{V}, \mathcal{E})$ is a directed graph, then a node X_i is a *parent* of node X_j if $(X_i, X_j) \in \mathcal{E}$. Moreover, if X_i is parent of X_j, then X_j is said to be *child* of X_i.

Definition B.6. If $\mathcal{G} = (\mathcal{V}, \mathcal{E})$ is a directed graph, then a node X_i is a *root* of $\mathcal{G}$ if there does not exist any node X_j such that $(X_j, X_i) \in \mathcal{E}$ (i.e. if X_i has no parents in $\mathcal{G}$).

Another useful concept is the concept of *path*. First, we define paths for directed graphs:

Definition B.7. If $\mathcal{G} = (\mathcal{V}, \mathcal{E})$ is a directed graph, then a *path* $\mathcal{P}$ from X_i to X_j (where $\{X_i, X_j\} \subseteq \mathcal{V}$) is a subset of $\mathcal{E}$ satisfying the following conditions:

1. $\exists k\ (X_i, X_k) \in \mathcal{P}$;
2. $\exists k\ (X_k, X_j) \in \mathcal{P}$;
3. Any edge (X_k, X_l) in $\mathcal{P}$ displays the following properties:

   a. If $k \neq i$, then $\exists m\ (X_m, X_k) \in \mathcal{P}$;
   b. If $l \neq j$, then $\exists m\ (X_l, X_m) \in \mathcal{P}$.

Given the definition of path, we can define the notions of *ancestor* and *descendant*:

Definition B.8. Given a directed graph $\mathcal{G} = (\mathcal{V}, \mathcal{E})$, if $\mathcal{P}$ is a *path* from X_i to X_j (where $\{X_i, X_j\} \subseteq \mathcal{V}$), then we say that X_i is an *ancestor* of X_j and X_j is a *descendant* of X_i.

On the other hand, we define paths for undirected graphs:

Definition B.9. If $\mathcal{G} = (\mathcal{V}, \mathcal{E})$ is an undirected graph, then a *path* $\mathcal{P}$ from X_i to X_j (where $\{X_i, X_j\} \subseteq \mathcal{V}$) is a subset of $\mathcal{E}$ satisfying the following conditions:

1. $\exists k\ \{X_i, X_k\} \in \mathcal{P}$;
2. $\exists k\ \{X_j, X_k\} \in \mathcal{P}$;
3. Any edge $\{X_k, X_l\}$ in $\mathcal{P}$ displays the following properties:

   a. If $k \notin \{i, j\}$, then $\exists m\ \{X_k, X_m\} \in \mathcal{P}$;
   b. If $l \notin \{i, j\}$, then $\exists m\ \{X_l, X_m\} \in \mathcal{P}$.

More sophisticated notions can be defined for directed graphs. First, we introduce *chains*:

Definition B.10. If $\mathcal{G} = (\mathcal{V}, \mathcal{E})$ is a directed graph, then a *chain* $\mathcal{C}$ from X_i to X_j (where $\{X_i, X_j\} \subseteq \mathcal{V}$) is a subset of $\mathcal{E}$ satisfying the following conditions:

1. $\exists k\ (X_i, X_k) \in \mathcal{C} \vee (X_k, X_i) \in \mathcal{C}$;
2. $\exists k\ (X_j, X_k) \in \mathcal{C} \vee (X_k, X_j) \in \mathcal{C}$;
3. Any edge (X_k, X_l) in $\mathcal{C}$ displays the following properties:

   a. If $k \notin \{i, j\}$, then $\exists m\ (X_k, X_m) \in \mathcal{C} \vee (X_m, X_k) \in \mathcal{C}$;
   b. If $l \notin \{i, j\}$, then $\exists m\ (X_l, X_m) \in \mathcal{C} \vee (X_m, X_l) \in \mathcal{C}$.

If a set S of pairs $\{X_1,X_2\},\ldots,\{X_{n-1},X_n\}$ is such that $X_i \in \{X_1,X_2\}$, $X_j \in \{X_{n-1},X_n\}$, and (for $2 \le i \le |\mathcal{V}|$) $\{X_{i-1},X_i\} \in S \Leftrightarrow ((X_{i-1},X_i) \in \mathcal{C} \vee (X_i,X_{i-1}) \in \mathcal{C})$, then $\mathcal{C}$ is denoted equivalently by $[X_1,\ldots,X_n]$ and $[X_n,\ldots,X_1]$.

Informally, we often say that a chain $\mathcal{C}$ contains a node X to mean that $\mathcal{C}$ contains an edge (Y,Z) such that $X = Y$ or $X = Z$.

Second, we characterize *uncoupled meetings*:

Definition B.11. An *uncoupled meeting* is a chain $[X_i,X_j,X_k]$ such that X_i and X_k are not adjacent.

Third, we define *directed* and *undirected cycles* respectively:

Definition B.12. If $\mathcal{G}$ is a directed graph, then an *undirected cycle* in $\mathcal{G}$ is a chain from a node to itself.

Definition B.13. If $\mathcal{G}$ is a directed graph, then a *directed cycle* in $\mathcal{G}$ is a path from a node to itself.

We now define a directed acyclic graph (DAG):

Definition B.14. A directed graph $\mathcal{G}$ is *acyclic* if $\mathcal{G}$ does not contain any directed cycle.

Finally, we specify the concept of *topological* (or *ancestral*) *ordering* for directed acyclic graphs:

Definition B.15. If $\mathcal{G} = (\mathcal{V},\mathcal{E})$ is a DAG containing n nodes, then a *topological* (or *ancestral*) *ordering* $(X_1,\ldots,X_n)$ of the nodes in $\mathcal{V}$ is any ordering such that, if $(X_i,X_j) \in \mathcal{E}$, then $i < j$.

B.2 Algorithms

Two important problems usually encountered while working with directed graphs are given on the one hand by the need to verify whether the graph is acyclic, on the other hand by the task of sorting the nodes in a topological ordering. We now describe an algorithm (first introduced by Arthur Kahn [157]) which is suitable for solving both problems at the same time. The algorithm takes as input a directed graph $\mathcal{G}$. Two lists are defined, S and $\mathcal{U}$, containing sorted and unsorted nodes respectively. $\mathcal{U}$ is initialized as the list of the roots (if any) of $\mathcal{G}$. While $\mathcal{U}$ is not empty, one node X_i is removed from it and appended to S. Then, for each child X_j of X_i, we perform the following two steps. First, (X_i,X_j) is removed from $\mathcal{E}$. Second, if $\mathcal{E}$ does not contain any other edge (X_k,X_j), then X_j is added to $\mathcal{U}$. Once the while loop terminates, the following condition is checked in order to return the proper output. If $\mathcal{E}$ is empty, then the graph is acyclic and the elements of S are sorted according to a topological ordering, otherwise the graph contains at least one (directed) cycle and no topological ordering is possible over $\mathcal{V}$. Pseudocode for the described technique is provided by Algorithm B.1.

Algorithm B.1 Kahn: Topological sorting of nodes in directed acyclic graphs

---

Input: Directed graph $\mathcal{G} = (\mathcal{V}, \mathcal{E})$.
Output: Ordered set $\mathcal{S}$: if $\mathcal{S} = \emptyset$, then $\mathcal{G}$ is not acyclic, otherwise $\mathcal{S}$ provides a topological ordering over $\mathcal{V}$.

Kahn($\mathcal{G}$):
```
 1.    S = (S₁,...,S|ᵥ|)
 2.    U = {Xᵢ : Xᵢ ∈ V ∧ ∄j (Xⱼ,Xᵢ) ∈ E}
 3.    E* = E
 4.    i = 1
 5.    while(U ≠ ∅)
 6.        Xⱼ = an arbitrary element of U
 7.        U = U \ {Xⱼ}
 8.        Sᵢ = Xⱼ
 9.        i = i + 1
10.        for(Xₖ ∈ {Xₗ : (Xⱼ,Xₗ) ∈ E})
11.            E* = E* \ {(Xⱼ,Xₖ)}
12.            if(∄l (Xₗ,Xₖ) ∈ E*)
13.                U = U ∪ {Xₖ}
14.        if(E* ≠ ∅)
15.            S = ∅
16.    return S
```

1. $\mathcal{S} = (S_1, \ldots, S_{|\mathcal{V}|})$
2. $\mathcal{U} = \{X_i : X_i \in \mathcal{V} \wedge \nexists j \, (X_j, X_i) \in \mathcal{E}\}$
3. $\mathcal{E}^* = \mathcal{E}$
4. $i = 1$
5. $\texttt{while}(\mathcal{U} \neq \emptyset)$
6. $X_j = $ an arbitrary element of $\mathcal{U}$
7. $\mathcal{U} = \mathcal{U} \setminus \{X_j\}$
8. $S_i = X_j$
9. $i = i + 1$
10. $\texttt{for}(X_k \in \{X_l : (X_j, X_l) \in \mathcal{E}\})$
11. $\mathcal{E}^* = \mathcal{E}^* \setminus \{(X_j, X_k)\}$
12. $\texttt{if}(\nexists l \, (X_l, X_k) \in \mathcal{E}^*)$
13. $\mathcal{U} = \mathcal{U} \cup \{X_k\}$
14. $\texttt{if}(\mathcal{E}^* \neq \emptyset)$
15. $\mathcal{S} = \emptyset$
16. $\texttt{return } \mathcal{S}$

When working with an undirected graph $\mathcal{G}$, it may be necessary to find all maximal cliques contained in $\mathcal{G}$. Algorithm B.2 provides pseudocode for the Bron-Kerbosch method, first introduced in [37], which is known to be one of the most efficient methods for finding all maximal cliques in undirected graphs. Actually, Algorithm B.2 presents a variant of the Bron-Kerbosch method using the so-called *pivoting* technique. Pivoting consists in using node X_i at line 3 as a 'pivot' for restricting the recursive call of line 6 to those nodes in $\mathcal{V}^*$ that are not adjacent to X_i. On the other hand, the version of the Bron-Kerbosch algorithm that does not use pivoting would iterate the instructions at lines 6–8 for all elements of $\mathcal{V}^*$. This strategy turns out to be much less efficient than the variant with pivoting for graphs containing a large number of non-maximal cliques. A theoretical and experimental analysis of the Bron-Kerbosch algorithm is offered in [37].

Algorithm B.2 BronKerbosch: Finding all maximal cliques in undirected graphs

---

Input: Undirected graph $\mathcal{G} = (\mathcal{V}, \mathcal{E})$; sets $\mathcal{C} = \emptyset$, $\mathcal{V}^* = \mathcal{V}$, $\mathcal{X} = \emptyset$, $\gamma = \emptyset$.
Output: Set γ containing all maximal cliques of $\mathcal{G}$.

BronKerbosch$(\mathcal{G}, \mathcal{C}, \mathcal{V}^*, \mathcal{X}, \gamma)$:
1.   if $(\mathcal{V}^* \cup \mathcal{X} = \emptyset)$
2. $\gamma = \gamma \cup \{\mathcal{C}\}$
3. X_i = an arbitrary element of $\mathcal{V}^* \cup \mathcal{X}$
4. $\mathcal{N} = \{X_j : \{X_i, X_j\} \in \mathcal{E}\}$
5.   for $(X_j \in \mathcal{V}^* \setminus \mathcal{N})$
6. $\gamma = $ BronKerbosch$(\mathcal{G}, \mathcal{C} \cup \{X_j\}, \mathcal{V}^* \cap \mathcal{N}, \mathcal{X} \cap \mathcal{N}, \gamma))$
7. $\mathcal{V}^* = \mathcal{V}^* \setminus \{X_j\}$
8. $\mathcal{X} = \mathcal{X} \cup \{X_j\}$
9.   return γ

Afterword

Aside from the tutorial parts of this volume, where we tried to render the text as plain and self-contained as possible, this is mostly a research book. In this perspective, our attempt has been to provide the reader with fresh and up-to-date contents, including new ideas and recent developments of an on-going investigation field we are constantly involved in. Nonetheless, some portions of the book were previously, yet partially released in the form of journal articles or contributions to conference proceedings. This has been made clear explicitly throughout the text by proper references to the earlier bibliographic sources. Episodic reproductions of published contents strictly comply with the original Publishers' copyright policies we subscribed to. In particular, some material drawn from papers published by either ACM[1] [94], Elsevier[2] [94], or IOS[3] [97] partially appears within the following sections of the book: 4.6, 5.5.1–5.5.2, 5.6, 6.3–6.4.

Siena,   Antonino Freno
January 2011   Edmondo Trentin

---

[1] http://www.acm.org/.

[2] http://www.elsevier.com/.

[3] http://www.iospress.nl/.

References

1. Ackley, D.H., Hinton, G.E., Sejnowski, T.J.: A Learning Algorithm for Boltzmann Machines. Cognitive Science 9(1), 147–169 (1985)
2. Adler, R.J., Taylor, J.E.: Random Fields and Geometry. Springer, New York (2007)
3. Aha, D.W., Kibler, D., Albert, M.K.: Instance-Based Learning Algorithms. Machine Learning 6, 37–66 (1991)
4. Akaike, H.: A new look at the statistical model identification. IEEE Transactions on Automatic Control 19, 716–723 (1974)
5. Albert, R., Barabási, A.L.: Topology of evolving networks: Local events and universality. Physical Review Letters 85, 5234–5237 (2000)
6. Aliferis, C.F., Statnikov, A., Tsamardinos, I., Mani, S., Koutsoukos, X.D.: Local Causal and Markov Blanket Induction for Causal Discovery and Feature Selection for Classification. Part I: Algorithms and Empirical Evaluation. Journal of Machine Learning Research, 171–234 (2010)
7. Allcroft, D.J., Glasbey, C.A.: A Latent Gaussian Markov Random-Field Model for Spatiotemporal Rainfall Disaggregation. Journal of the Royal Statistical Society. Series C (Applied Statistics) 52(4), 487–498 (2003)
8. Ash, R.B.: Basic Probability Theory, 2nd edn. Dover Publications, Mineola (2008)
9. Bacchus, F.: Representing and Reasoning with Probabilistic Knowledge: A Logical Approach to Probabilities. MIT Press, Cambridge (1990)
10. Bach, F.R., Jordan, M.I.: Learning Graphical Models with Mercer Kernels. In: Advances in Neural Information Processing Systems, pp. 1009–1016 (2002)
11. Baker, A.: Simplicity. In: Zalta, E.N. (ed.) The Stanford Encyclopedia of Philosophy, Stanford University, Stanford (2004),
    http://plato.stanford.edu/archives/win2004/entries/
    simplicity/
12. Baldi, P., Brunak, S.: Bioinformatics: The Machine Learning Approach, 2nd edn. MIT Press, Cambridge (2001)
13. Barabási, A.L., Albert, R.: Emergence of scaling in random networks. Science 286, 509–512 (1999)
14. Barnes, E.C.: Ockham's Razor and the Anti-Superfluity Principle. Erkenntnis 53, 353–374 (2000)
15. Bartlett, M.S.: Contingency Table Interaction. Supplement to the Journal of the Royal Statistical Society 2, 248–252 (1935)
16. Basener, W.F.: Topology and Its Applications. John Wiley & Sons, New York (2006)

17. Bayes, T.: An Essay towards solving a Problem in the Doctrine of Chances. By the late Rev. Mr. Bayes, F. R. S. Communicated by Mr. Price, in a letter to John Canton, A. M. F. R. S. Philosophical Transactions of the Royal Society of London 53, 370–418 (1763)

18. Bechtel, W.: Philosophy of Mind: An Overview for Cognitive Science. Lawrence Erlbaum Associates, Hillsdale (1988)

19. Bechtel, W.: Philosophy of Science: An Overview for Cognitive Science. Lawrence Erlbaum Associates, Hillsdale (1988)

20. Beeri, C., Fagin, R., Maier, D., Yannakakis, M.: On the Desirability of Acyclic Database Schemes. Journal of the ACM 30, 479–513 (1983)

21. Bengio, Y., Frasconi, P.: Input-output hmm's for sequence processing. IEEE Transactions on Neural Networks 7(5), 1231–1249 (1996)

22. Benjafield, J.G.: Cognition, 2nd edn. Prentice Hall, Upper Saddle River (1997)

23. Bentley, J.L.: Multidimensional Binary Search Trees Used for Associative Searching. Communications of the ACM 18, 509–517 (1975)

24. Besag, J.: Spatial Interaction and the Statistical Analysis of Lattice Systems. Journal of the Royal Statistical Society Series B 36, 192–236 (1974)

25. Besag, J.: Statistical Analysis of Non-Lattice Data. The Statistician 24, 179–195 (1975)

26. Besag, J., Green, P., Higdon, D., Mengersen, K.: Bayesian computation and stochastic systems. Statistical Sciences 10, 3–66 (1995)

27. Besag, J., Kooperberg, C.: On Conditional and Intrinsic Autoregressions. Biometrika 82, 733–746 (1995)

28. Bickle, J.W.: Psychoneural Reduction. MIT Press, Cambridge (1998)

29. Bishop, C.M.: Neural Networks for Pattern Recognition. Oxford University Press, Oxford (1995)

30. Bishop, C.M.: Pattern Recognition and Machine Learning. Springer, New York (2006)

31. Bourlard, H., Morgan, N.: Connectionist Speech Recognition. A Hybrid Approach. Kluwer Academic Publishers, Boston (1994)

32. de Salvo Braz, R., Amir, E., Roth, D.: A survey of first-order probabilistic models. In: Holmes, D.E., Jain, L.C. (eds.) Innovations in Bayesian Networks: Theory and Applications, pp. 289–317. Springer, Berlin (2008)

33. Breese, J.S., Heckerman, D., Kadie, C.M.: Empirical Analysis of Predictive Algorithms for Collaborative Filtering. In: Proceedings of the Fourteenth Annual Conference on Uncertainty in Artificial Intelligence, pp. 43–52 (1998)

34. Brezger, A., Fahrmeir, L., Hennerfeind, A.: Adaptive Gaussian Markov random fields with applications in human brain mapping. Journal of the Royal Statistical Society. Series C (Applied Statistics) 56, 327–345 (2007)

35. Brillinger, D.R.: A Potential Function Approach to the Flow of Play in Soccer. Journal of Quantitative Analysis in Sports 3(1) (2007)

36. Brin, S., Page, L.: The Anatomy of a Large-Scale Hypertextual Web Search Engine. In: Proceedings of the 7th International World Wide Web Conference (WWW7). Computer Networks, vol. 30, pp. 107–117 (1998)

37. Bron, C., Kerbosch, J.: Finding All Cliques of an Undirected Graph. Communications of the ACM 16, 575–577 (1973)

38. Buntine, W.L.: Theory Refinement on Bayesian Networks. In: Proceedings of the Seventh Annual Conference on Uncertainty in Artificial Intelligence (UAI 1991), pp. 52–60. Morgan Kaufmann, San Francisco (1991)

39. Buntine, W.L.: Operations for Learning with Graphical Models. Journal of Artificial Intelligence Research 2, 159–225 (1994)

40. Buntine, W.L.: Chain Graphs for Learning. In: UAI 1995: Proceedings of the Eleventh Annual Conference on Uncertainty in Artificial Intelligence, pp. 46–54. Morgan Kaufmann, San Francisco (1995)
41. de Campos, L.M., Romero, A.E.: Bayesian network models for hierarchical text classification from a thesaurus. International Journal of Approximate Reasoning 50(7), 932–944 (2009)
42. Carroll, S., Pavlovic, V.: Protein Classification Using Probabilistic Chain Graphs and the Gene Ontology Structure. Bioinformatics 22, 1871–1878 (2006)
43. Catlett, J.: On changing continuous attributes into ordered discrete attributes. In: Kodratoff, Y. (ed.) EWSL 1991. LNCS, vol. 482, pp. 164–178. Springer, Heidelberg (1991)
44. Ceroni, A., Costa, F., Frasconi, P.: Classification of small molecules by two- and three-dimensional decomposition kernels. Bioinformatics 23(16), 2038–2045 (2007)
45. Chandler, D.: Introduction to Modern Statistical Mechanics. Oxford University Press, New York (1987)
46. Chang, K.C., Fung, R.M.: Refinement and coarsening of bayesian networks. In: Proceedings of the Sixth Annual Conference on Uncertainty in Artificial Intelligence (UAI 1990), pp. 435–446 (1990)
47. Chellappa, R., Jain, A.K. (eds.): Markov Random Fields: Theory and Applications. Academic Press, Boston (1993)
48. Churchland, P.M.: Eliminative Materialism and the Propositional Attitudes. The Journal of Philosophy 78, 67–90 (1981)
49. Cohen, J.: Bioinformatics—An Introduction for Computer Scientists. ACM Computing Surveys 36, 122–158 (2004)
50. Cooper, G.F., Herskovits, E.: A Bayesian Method for the Induction of Probabilistic Networks from Data. Machine Learning 9, 309–347 (1992)
51. Cormen, T.H., Leiserson, C.E., Rivest, R.L., Stein, C.: Introduction to Algorithms. MIT Press, Cambridge (2001)
52. Cowell, R.G.: Local Propagation in Conditional Gaussian Bayesian Networks. Journal of Machine Learning Research 6, 1517–1550 (2005)
53. Cowell, R.G., Dawid, A.P., Lauritzen, S.L., Spiegelhalter, D.J.: Probabilistic Networks and Expert Systems. In: Exact Computational Methods for Bayesian Networks. Springer, New York (1999)
54. Crane, T.: The Mechanical Mind, 2nd edn. Routledge, London (2003) (First published 1995)
55. Cressie, N.A.C.: Statistics for Spatial Data, 2nd edn. John Wiley & Sons, New York (1993)
56. van Dalen, D.: Logic and Structure, 3rd edn. Springer, Berlin (1997)
57. Dawid, A.P., Lauritzen, S.L.: Hyper Markov Laws in the Statistical Analysis of Decomposable Graphical Models. The Annals of Statistics 21(3), 1272–1317 (1993)
58. De Mori, R. (ed.): Spoken Dialogues with Computers. Academic Press, San Diego (1998)
59. De Raedt, L., Dehaspe, L.: Clausal Discovery. Machine Learning 26, 99–146 (1997)
60. De Raedt, L., Dietterich, T.G., Getoor, L., Muggleton, S.H. (eds.): Probabilistic, Logical and Relational Learning – Towards a Synthesis. Internationales Begegnungs- und Forschungszentrum für Informatik (IBFI), Schloss Dagstuhl, Germany, Dagstuhl, Germany (2006)
61. D'Elia, C., Poggi, G., Scarpa, G.: A tree-structured Markov random field model for Bayesian image segmentation. IEEE Transactions on Image Processing 12(10), 1259–1273 (2003)

62. Dempster, A.P., Laird, N.M., Rubin, D.B.: Maximum Likelihood from Incomplete Data via the EM Algorithm. Journal of the Royal Statistical Society Series B 39, 1–38 (1977)

63. Denoyer, L., Gallinari, P.: Bayesian network model for semi-structured document classification. Information Processing and Management 40(5), 807–827 (2004)

64. Derin, H., Elliott, H., Cristi, R., Geman, D.: Bayes Smoothing Algorithms for Segmentation of Binary Images Modeled by Markov Random Fields. IEEE Transactions on Pattern Analysis and Machine Intelligence 6, 707–720 (1984)

65. Devroye, L.: A Course in Density Estimation. Birkhauser Boston Inc., Cambridge (1987)

66. Diebel, J., Thrun, S.: An Application of Markov Random Fields to Range Sensing. In: Advances in Neural Information Processing Systems (NIPS), pp. 291–298. MIT Press, Cambridge (2005)

67. Diestel, R.: Graph Theory, 3rd edn. Springer, Berlin (2006)

68. Dietterich, T.G., Getoor, L., Murphy, K. (eds.): Proceedings of the ICML 2004 Workshop on Statistical Relational Learning and its Connections to Other Fields (SRL 2004), IMLS, Banff Canada (2004)

69. Dixon, W.J.: Simplified Estimation from Censored Normal Samples. The Annals of Mathematical Statistics 31(2), 385–391 (1960)

70. Domingos, P., Pazzani, M.: On the Optimality of the Simple Bayesian Classifier under Zero-One Loss. Machine Learning 29, 103–130 (1997)

71. Doob, J.L.: Stochastic Processes. John Wiley & Sons, New York (1953)

72. Duda, R.O., Hart, P.E.: Pattern Classification and Scene Analysis. John Wiley & Sons, New York (1973)

73. Duda, R.O., Hart, P.E., Stork, D.G.: Pattern Classification, 2nd edn. John Wiley & Sons, New York (2001)

74. Durbin, R., Eddy, S., Krogh, A., Mitchison, G.: Biological Sequence Analysis. Probabilistic models of proteins and Nucleic Acids. Cambridge University Press, Cambridge (1998)

75. Ebbinghaus, H.D., Flum, J.: Finite Model Theory, 2nd edn. Springer, Berlin (1999)

76. Efron, B.: Better Bootstrap Confidence Intervals. Journal of the American Statistical Association 82, 171–185 (1981)

77. Efron, B.: Nonparametric Estimates of Standard Error: The Jackknife, the Bootstrap and Other Methods. Biometrika 68, 589–599 (1981)

78. Efron, B., Tibshirani, R.J.: An Introduction to the Bootstrap. Chapman & Hall/CRC, Boca Raton, FL (1993)

79. Elman, J.L.: Finding structure in time. Cognitive Science 14(2), 179–211 (1990)

80. Epanechnikov, V.: Nonparametric Estimation of a Multidimensional Probability Density. Theory of Probability and its Applications 14, 153–158 (1969)

81. Erdös, P., Rényi, A.: On random graphs. Publicationes Mathematicae Debrecen 6, 290–297 (1959)

82. Evans, M., Hastings, N., Peacock, B.: Statistical Distributions, third edn. John Wiley & Sons, New York (2000)

83. Faugeras, O.P.: A Quantile-Copula Approach to Conditional Density Estimation. Journal of Multivariate Analysis 100, 2083–2099 (2009)

84. Fausett, L.V.: Fundamentals of Neural Networks: Architectures, Algorithms, and Applications. Prentice Hall, Englewood Cliffs (1994)

85. Fayyad, U.M., Irani, K.B.: Multi-interval discretization of continuous-valued attributes for classification learning. In: Proceedings of the 13th International Joint Conference on Artificial Intelligence (IJCAI), pp. 1022–1027 (1993)

86. Feller, W.: An Introduction to Probability Theory and Its Applications, 3rd edn., vol. 1. John Wiley & Sons, New York (1968)
87. Feller, W.: An Introduction to Probability Theory and Its Applications, 2nd edn., vol. 2. John Wiley & Sons, New York (1971)
88. Feng, W., Jia, J., Liu, Z.Q.: Self-Validated Labeling of Markov Random Fields for Image Segmentation. IEEE Transactions on Pattern Analysis and Machine Intelligence 32(10), 1871–1887 (2010)
89. Fernández, A., Salmerón, A.: Extension of bayesian network classifiers to regression problems. In: Geffner, H., Prada, R., Machado Alexandre, I., David, N. (eds.) IBERAMIA 2008. LNCS (LNAI), vol. 5290, pp. 83–92. Springer, Heidelberg (2008)
90. Flury, B.: A First Course in Multivariate Statistics. Springer, New York (1997)
91. Forbes, F., Peyrard, N.: Hidden markov random field model selection criteria based on mean field-like approximations. IEEE Transactions on Pattern Analysis and Machine Intelligence 25, 1089–1101 (2003)
92. Forster, M., Sober, E.: How to Tell when Simpler, More Unified, or Less *Ad Hoc* Theories will Provide More Accurate Predictions. British Journal for the Philosophy of Science 45, 1–35 (1994)
93. Freno, A.: Selecting features by learning markov blankets. In: Apolloni, B., Howlett, R.J., Jain, L.C. (eds.) KES 2007, Part I. LNCS (LNAI), vol. 4692, pp. 69–76. Springer, Heidelberg (2007)
94. Freno, A., Trentin, E., Gori, M.: A Hybrid Random Field Model for Scalable Statistical Learning. Neural Networks 22, 603–613 (2009)
95. Freno, A., Trentin, E., Gori, M.: Scalable Pseudo-Likelihood Estimation in Hybrid Random Fields. In: Elder, J., Fogelman-Souli, F., Flach, P., Zaki, M. (eds.) Proceedings of the 15th ACM SIGKDD Conference on Knowledge Discovery and Data Mining (KDD 2009), pp. 319–327. ACM, New York (2009)
96. Freno, A., Trentin, E., Gori, M.: Scalable Statistical Learning: A Modular Bayesian/Markov Network Approach. In: Proceedings of the International Joint Conference on Neural Networks (IJCNN 2009), pp. 890–897. IEEE, Los Alamitos (2009)
97. Freno, A., Trentin, E., Gori, M.: Kernel-Based Hybrid Random Fields for Nonparametric Density Estimation. In: 19th European Conference on Artificial Intelligence (ECAI 2010), pp. 427–432. IOS Press, Amsterdam (2010)
98. Friedman, J., Hastie, T., Tibshirani, R.: Sparse Inverse Covariance Estimation with the Graphical Lasso. Biostatistics 9, 432–441 (2008)
99. Friedman, N., Getoor, L., Koller, D., Pfeffer, A.: Learning Probabilistic Relational Models. In: Proceedings of the Sixteenth International Joint Conference on Artificial Intelligence (IJCAI 1999), pp. 1300–1309. Morgan Kaufmann, Stockholm (1999)
100. Friedman, N., Goldszmidt, M.: Learning Bayesian Networks with Local Structure. In: Proceedings of the Twelfth Annual Conference on Uncertainty in Artificial Intelligence (UAI 1996), pp. 252–262. Morgan Kaufmann, San Francisco (1996)
101. Friedman, N., Koller, D.: Being Bayesian about Bayesian Network Structure: A Bayesian Approach to Structure Discovery in Bayesian Networks. Machine Learning 50, 95–125 (2003)
102. Friedman, N., Linial, M., Nachman, I., Pe'er, D.: Using bayesian networks to analyze expression data. Journal of Computational Biology 7, 601–620 (2000)
103. Frydenberg, M., Lauritzen, S.L.: Decomposition of Maximum Likelihood in Mixed Graphical Interaction Models. Biometrika 76(3), 539–555 (1989)
104. Fukunaga, K.: Statistical Pattern Recognition, 2nd edn. Academic Press, San Diego (1990)

105. Gärtner, T.: A survey of kernels for structured data. SIGKDD Explorations 5(1), 49–58 (2003)
106. Geman, S., Geman, D.: Stochastic Relaxation, Gibbs Distributions, and the Bayesian Restoration of Images. IEEE Transactions on Pattern Analysis and Machine Intelligence 6, 721–741 (1984)
107. Getoor, L., Friedman, N., Koller, D., Taskar, B.: Learning Probabilistic Models of Link Structure. Journal of Machine Learning Research 3, 679–707 (2002)
108. Getoor, L., Jensen, D. (eds.): Proceedings of the AAAI-2000 Workshop on Learning Statistical Models from Relational Data. AAAI Press, Austin (2000)
109. Getoor, L., Taskar, B. (eds.): Introduction to Statistical Relational Learning. MIT Press, Cambridge (2007)
110. Ghahramani, Z.: Learning dynamic bayesian networks. In: Giles, C.L., Gori, M. (eds.) IIASS-EMFCSC-School 1997. LNCS (LNAI), vol. 1387, pp. 168–197. Springer, Heidelberg (1998)
111. Gibbs, J.W.: Elementary Principles in Statistical Mechanics: Developed with Especial Reference to the Rational Foundation of Thermodynamics. Charles Scribner's Sons, New York (1902)
112. Gilbert, E.N.: Random graphs. Annals of Mathematical Statistics 30, 1141–1144 (1959)
113. Gilks, W.R., Richardson, S., Spiegelhalter, D.: Markov Chain Monte Carlo in Practice. Chapman & Hall/CRC (1996)
114. Glover, F., Laguna, M.: Tabu Search. Kluwer Academic Publishers, Boston (1997)
115. Gómez, V., Kappen, H.J., Chertkov, M.: Approximate Inference on Planar Graphs using Loop Calculus and Belief Propagation. Journal of Machine Learning Research 11, 1273–1296 (2010)
116. Gori, M., Monfardini, G., Scarselli, F.: A new model for learning in graph domains. In: Proceedings of the International Joint Conference on Neural Networks, IJCNN 2005 (2005)
117. Gori, M., Pucci, A.: ItemRank: A Random-Walk Based Scoring Algorithm for Recommender Engines. In: 20th International Joint Conference on Artificial Intelligence (IJCAI 2007), pp. 2766–2771 (2007)
118. Grädel, E., Kolaitis, P.G., Libkin, L., Marx, M., Spencer, J., Vardi, M.Y., Venema, Y., Weinstein, S.: Finite Model Theory and Its Applications. Springer, Heidelberg (2007)
119. Gravier, G., Sigelle, M., Chollet, G.: A markov random field model for automatic speech recognition. In: Proceedings of the International Conference on Pattern Recognition, pp. 3258–3261 (2000)
120. Gray, A.G., Moore, A.W.: 'N-Body' Problems in Statistical Learning. In: Advances in Neural Information Processing Systems, pp. 521–527 (2000)
121. Griffeath, D.: Introduction to Random Fields. In: Kemeny, J.G., Snell, J.L., Knapp, A.W. (eds.) Denumerable Markov Chains, 2nd edn., ch. 12. Springer, New York (1976)
122. Guyon, I., Elisseeff, A.: An Introduction to Variable and Feature Selection. Journal of Machine Learning Research 3, 1157–1182 (2003)
123. Hall, M.A.: Correlation-based Feature Selection for Discrete and Numeric Class Machine Learning. In: Proceedings of the Seventeenth International Conference on Machine Learning (ICML 2000), pp. 359–366. Morgan Kaufmann, San Francisco (2000)
124. Hammer, B., Micheli, A., Sperduti, A., Strickert, M.: Recursive self-organizing network models. Neural Networks 17, 1061–1085 (2004)
125. Hammer, B., Saunders, C., Sperduti, A.: Special issue on neural networks and kernel methods for structured domains. Neural Networks 18(8), 1015–1018 (2005)
126. Hand, D.J., Yu, K.: Idiot's Bayes—Not So Stupid After All? International Statistical Review 69, 385–398 (2001)

127. Haykin, S.: Neural Networks. A Comprehensive Foundation, 2nd edn. Prentice Hall, New York (1999)
128. Heckerman, D., Chickering, D.M., Meek, C., Rounthwaite, R., Kadie, C.M.: Dependency Networks for Inference, Collaborative Filtering, and Data Visualization. Journal of Machine Learning Research 1, 49–75 (2000)
129. Heckerman, D., Geiger, D., Chickering, D.M.: Learning Bayesian Networks: The Combination of Knowledge and Statistical Data. Machine Learning 20, 197–243 (1995)
130. Held, K., Kops, E.R., Krause, B.J., Wells, W.M.I., Kikinis, R., Müller-Gärtner, H.W.: Markov random field segmentation of brain MR images. IEEE Transactions on Medical Imaging 16(6), 878–886 (1997)
131. Helman, P., Veroff, R., Atlas, S.R., Willman, C.: A Bayesian Network Classification Methodology for Gene Expression Data. Journal of Computational Biology 11(4), 581–615 (2004)
132. Herskovits, E., Cooper, G.: Kutató: An Entropy-Driven System for Construction of Probabilistic Expert Systems from Databases. In: Bonissone, P., Henrion, M., Kanal, L.N., Lemmer, J.F. (eds.) Proceedings of the Sixth Conference Annual Conference on Uncertainty in Artificial Intelligence (UAI 1990), pp. 54–63. Elsevier Science, New York (1990)
133. Hertz, J., Krogh, A., Palmer, R.G.: Introduction to the Theory of Neural Computation. Addison-Wesley, Redwood City (1991)
134. Hilbert, D.: Über das Unendliche. Mathematische Annalen 95, 161–190 (1926)
135. Hirsch, M.W.: Differential Topology. Springer, Heidelberg (1997)
136. Hofmann, R., Tresp, V.: Discovering Structure in Continous Variables Using Bayesian Networks. In: Advances in Neural Information Processing Systems, pp. 500–506 (1995)
137. Hofmann, R., Tresp, V.: Nonlinear Markov Networks for Continous Variables. In: Advances in Neural Information Processing Systems (1997)
138. Holmes, M.P., Gray, A.G., Isbell, C.L.: Fast Nonparametric Conditional Density Estimation. In: Proceedings of the 23rd Conference on Uncertainty in Artificial Intelligence, UAI 2007 (2007)
139. Holmes, M.P., Gray, A.G., Isbell, C.L.J.: Fast Kernel Conditional Density Estimation: A Dual-Tree Monte Carlo Approach. Computational Statistics and Data Analysis 54, 1707–1718 (2010)
140. Hong, Z.Q., Yang, J.Y.: Optimal Discriminant Plane for a Small Number of Samples and Design Method of Classifier on the Plane. Pattern Recognition 24, 317–324 (1991)
141. Hopfield, J.J.: Neural networks and physical systems with emergent collective computational abilities. Proceedings of the National Academy of Sciences of the United States of America 79(8), 2554–2558 (1982)
142. Hrbacek, K., Jech, T.J.: Introduction to Set Theory, 3rd edn. Marcel Dekker, New York (1999)
143. Hruschka, E.R.J., Hruschka, E.R., Ebecken, N.F.F.: Feature Selection by Bayesian Networks. In: Tawfik, A.Y., Goodwin, S.D. (eds.) Canadian AI 2004. LNCS (LNAI), vol. 3060, pp. 370–379. Springer, Heidelberg (2004)
144. Hu, J., Brown, M.K., Turin, W.: HMM Based On-Line Handwriting Recognition. IEEE Transactions on Pattern Analysis and Machine Intelligence 18, 1039–1045 (1996)
145. Huang, K.: Statistical Mechanics. John Wiley & Sons, New York (1987)
146. Huber, P.J.: Robust Statistics. Wiley & Sons, Chichester (1981)
147. Hume, D.: A Treatise of Human Nature. Clarendon Press, Oxford (1896) (Original work published 1739)
148. Hwang, K.-B., Lee, J.-W., Chung, S.-W., Zhang, B.-T.: Construction of Large-Scale Bayesian Networks by Local to Global Search. In: Ishizuka, M., Sattar, A. (eds.) PRICAI 2002. LNCS (LNAI), vol. 2417, pp. 375–384. Springer, Heidelberg (2002)

149. Jaeger, M.: Relational Bayesian Networks. In: Proceedings of the Thirteenth Conference on Uncertainty in Artificial Intelligence (UAI 1997), pp. 266–273 (1997)
150. Jaeger, M.: Reasoning About Infinite Random Structures with Relational Bayesian Networks. In: Proceedings of the Sixth International Conference on Principles of Knowledge Representation and Reasoning, pp. 570–581 (1998)
151. Jain, A.K., Dubes, R.C.: Algorithms for Clustering Data. Prentice Hall, Englewood Cliffs (1988)
152. Jensen, F.V., Nielsen, T.D.: Bayesian Networks and Decision Graphs, 2nd edn. Springer, New York (2010)
153. John, G.H., Kohavi, R., Pfleger, K.: Irrelevant Features and the Subset Selection Problem. In: Cohen, W.W., Hirsh, H. (eds.) Machine Learning: Proceedings of the Eleventh International Conference, pp. 121–129. Morgan Kaufmann, San Francisco (1994)
154. Jordan, M.I. (ed.): Learning in Graphical Models. MIT Press, Cambridge (1999)
155. Jordan, M.I.: Graphical Models. Statistical Science 19, 140–155 (2004)
156. Jung, S.Y., Park, Y.C., Choi, K.S., Kim, Y.: Markov random field based English part-of-speech tagging system. In: Proceedings of the 16th conference on Computational linguistics (COLING 1996), vol. 1, pp. 236–242 (1996)
157. Kahn, A.B.: Topological Sorting of Large Networks. Communications of the ACM 5(11), 558–562 (1962)
158. Kalman, R.E.: A New Approach to Linear Filtering and Prediction Problems. Journal of Basic Engineering—Transactions of the ASME (Series D) 82, 35–45 (1960)
159. Kandel, E.R., Schwartz, J.H., Jessell, T.M. (eds.): Principles of Neural Science, 4th edn. McGraw-Hill, New York (2000)
160. Kaplan, E.L., Meier, P.: Nonparametric Estimation from Incomplete Observations. Journal of the American Statistical Association 53(282), 457–481 (1958)
161. Kaplan, W.: Advanced Calculus, 3rd edn. Addison-Wesley, Reading (1984)
162. Karp, R.: Reducibility among Combinatorial Problems. In: Miller, R.E., Thatcher, J.W. (eds.) Complexity of Computer Computations, pp. 85–103. Plenum Press, New York (1972)
163. Kashima, H., Koyanagi, T.: Kernels for semi-structured data. In: Sammut, C., Hoffmann, A.G. (eds.) Proceedings of the Nineteenth International Conference on Machine Learning (ICML 2002), pp. 291–298 (2002)
164. Kenney, J., Keeping, E.: Mathematics of Statistics. Part 2, 2nd edn. Van Nostrand, Princeton (1951)
165. Kenney, J., Keeping, E.: Mathematics of Statistics. Part 1, 3rd edn. Van Nostrand, Princeton (1962)
166. Kersting, K., De Raedt, L.: Towards Combining Inductive Logic Programming with Bayesian Networks. In: Rouveirol, C., Sebag, M. (eds.) ILP 2001. LNCS (LNAI), vol. 2157, pp. 118–131. Springer, Heidelberg (2001)
167. Kindermann, R., Snell, J.L.: Markov Random Fields and Their Applications. American Mathematical Society, Providence (1980)
168. Klemens, B.: Modeling with Data: Tools and Techniques for Scientific Computing. Princeton University Press, Princeton (2008)
169. Klopotek, M.A.: Very large bayesian multinets for text classification. Future Generation Computer Systems 21(7), 1068–1082 (2005)
170. Kohavi, R., John, G.H.: Wrappers for Feature Subset Selection. Artificial Intelligence 97, 273–324 (1997)
171. Kojima, K., Perrier, E., Imoto, S., Miyano, S.: Optimal Search on Clustered Structural Constraint for Learning Bayesian Network Structure. Journal of Machine Learning Research 11, 285–310 (2010)

172. Kok, S., Domingos, P.: Learning the Structure of Markov Logic Networks. In: Proceedings of the 22nd International Conference on Machine Learning (ICML 2005), Bonn, Germany (2005)

173. Kok, S., Domingos, P.: Learning markov logic network structure via hypergraph lifting. In: Proceedings of the 26th International Conference on Machine Learning, ICML 2009 (2009)

174. Koller, D., Friedman, N.: Probabilistic Graphical Models: Principles and Techniques. MIT Press, Cambridge (2009)

175. Koller, D., Pfeffer, A.: Object-Oriented Bayesian Networks. In: Proceedings of the Thirteenth Annual Conference on Uncertainty in AI (UAI), pp. 302–313 (1997)

176. Koller, D., Sahami, M.: Toward Optimal Feature Selection. In: Saitta, L. (ed.) Machine Learning: Proceedings of the Thirteenth International Conference, pp. 284–292. Morgan Kaufmann, San Francisco (1996)

177. Kolmogorov, A.N.: Grundbegriffe der Wahrscheinlichkeitsrechnung. Julius Springer, Berlin (1933)

178. Kolmogorov, A.N.: Zur Theorie der Markoffschen Ketten. Mathematische Annalen 112, 155–160 (1936)

179. Kornblith, H. (ed.): Naturalizing Epistemology, 2nd edn. MIT Press, Cambridge (1994)

180. Kschischang, F.R., Frey, B.J., Loeliger, H.A.: Factor Graphs and the Sum-Product Algorithm. IEEE Transactions on Information Theory 47, 498–519 (1998)

181. Kuhn, T.S.: The Structure of Scientific Revolutions, 3rd edn. University of Chicago Press, Chicago (1996) (Original work published 1962)

182. Kumar, M.P., Kolmogorov, V., Torr, P.H.: An Analysis of Convex Relaxations for MAP Estimation of Discrete MRFs. Journal of Machine Learning Research 10, 71–106 (2009)

183. Kumar, S., Hebert, M.: Discriminative Fields for Modeling Spatial Dependencies in Natural Images. In: Advances in Neural Information Processing Systems, NIPS 2003 (2004)

184. Kumar, V.P., Desai, U.B.: Image Interpretation Using Bayesian Networks. IEEE Transactions on Pattern Analysis and Machine Intelligence 18(1), 74–77 (1996)

185. Kunsch, H., Geman, S., Kehagias, A.: Hidden Markov Random Fields. The Annals of Applied Probability 5(3), 577–602 (1995)

186. Lafferty, J., McCallum, A., Pereira, F.: Conditional random fields: Probabilistic models for segmenting and labeling sequence data. In: Proceedings of the 18th International Conference on Machine Learning (ICML), pp. 282–289. Morgan Kaufmann, San Francisco (2001)

187. Langley, P., Simon, H.A., Bradshaw, G.L., Zytkow, J.M.: Scientific Discovery: Computational Explorations of the Creative Processes. MIT Press, Cambridge (1987)

188. Langseth, H., Bangsø, O.: Parameter Learning in Object-Oriented Bayesian Networks. Annals of Mathematics and Artificial Intelligence 32, 221–243 (2001)

189. Lauritzen, S.L.: Graphical Models. Clarendon Press, Oxford (1996)

190. Lauritzen, S.L., Spiegelhalter, D.J.: Local Computation and Probabilities on Graphical Structures and their Applications to Expert Systems. Journal of the Royal Statistical Society Series B 50(2), 157–224 (1988)

191. Lauritzen, S.L., Wermuth, N.: Graphical Models for Associations between Variables, some of which are Qualitative and some Quantitative. The Annals of Statistics 17, 31–57 (1989)

192. Lavrac, N., Dzeroski, S.: Inductive Logic Programming: Techniques and Applications. Ellis Horwood, New York (1994)

193. Lee, C.-H., Greiner, R., Schmidt, M.: Support Vector Random Fields for Spatial Classification. In: Jorge, A.M., Torgo, L., Brazdil, P.B., Camacho, R., Gama, J. (eds.) PKDD 2005. LNCS (LNAI), vol. 3721, pp. 121–132. Springer, Heidelberg (2005)

194. Lenz, W.: Beiträge zum Verständnis der magnetischen Eigenschaften in festen Körpern. Physikalische Zeitschrift 21, 613–615 (1920)

195. Li, S.Z.: Markov Random Field Modeling in Image Analysis. Springer, New York (2001)

196. Liu, D.C., Nocedal, J.: On the Limited Memory BFGS Method for Large Scale Optimization. Mathematical Programming 45, 503–528 (1989)

197. Liu, H., Lafferty, J., Wasserman, L.: The Nonparanormal: Semiparametric Estimation of High Dimensional Undirected Graphs. Journal of Machine Learning Research 10, 2295–2328 (2009)

198. Llorente, A., Manmatha, R., Rüger, S.M.: Image retrieval using Markov Random Fields and global image features. In: Proceedings of the ACM International Conference on Image and Video Retrieval, pp. 243–250 (2010)

199. Luettgen, M.R., Karl, W.C., Willsky, A.S., Tenney, R.R.: Multiscale representations of Markov random fields. In: IEEE International Conference on Acoustics, Speech, and Signal Processing, vol. 5, pp. 41–44 (1993)

200. Luis, R., Sucar, L.E., Morales, E.F.: Inductive Transfer for Learning Bayesian Networks. Machine Learning 79, 227–255 (2010)

201. Luo, J., Savakis, A.E., Singhal, A.: A bayesian network-based framework for semantic image understanding. Pattern Recognition 38, 919–934 (2005)

202. Lustgarten, J.L., Gopalakrishnan, V., Grover, H., Visweswaran, S.: Improving classification performance with discretization on biomedical datasets. In: Proceedings of the Fall Symposium of the American Medical Informatics Association, pp. 445–449 (2008)

203. Magdon-Ismail, M., Atiya, A.: Neural Networks for Density Estimation. In: Advances in Neural Information Processing Systems, pp. 522–528 (1995)

204. Maglogiannis, I., Vouyioukas, D., Aggelopoulos, C.: Face Detection and Recognition of Natural Human Emotion Using Markov Random Fields. Personal and Ubiquitous Computing 13(1), 95–101 (2009)

205. Mardia, K.V.: Multi-dimensional multivariate Gaussian Markov random fields with application to image processing. Journal of Multivariate Analysis 24, 265–284 (1988)

206. Margaritis, D.: Distribution-Free Learning of Bayesian Network Structure in Continuous Domains. In: AAAI, pp. 825–830 (2005)

207. Markov, A.A.: Rasprostranenie zakona bolshih chisel na velichiny, zavisyaschie drug ot druga. Izvestiya Fiziko-matematicheskogo obschestva pri Kazanskom universitete, 2-ya seriya 15(4), 135–156 (1906)

208. Massaro, D.W., Cohen, M.M.: Continuous versus discrete information processing in pattern recognition. Acta Psychologica 90(1-3), 193–209 (1995)

209. McCallum, A.: Efficiently inducing features of conditional random fields. In: Proceedings of the Nineteenth Annual Conference on Uncertainty in Artificial Intelligence, UAI 2003 (2003)

210. McLachlan, G.J.: Discriminant Analysis and Statistical Pattern Recognition. John Wiley & Sons, New York (1992)

211. McLachlan, G.J., Basford, K.E.: Mixture Models: Inference and Applications to Clustering. Marcel Dekker, New York (1988)

212. Mehlhorn, K., Sanders, P.: Algorithms and Data Structures: The Basic Toolbox. Springer, Berlin (2008)

213. Mengshoel, O.J., Wilkins, D.C.: Abstraction and aggregation in belief networks. In: Abstractions, Decisions, and Uncertainty: Papers from the AAAI Workshop, pp. 53–58. AAAI Press, Menlo Park (1997) Techical Report WS-97-08

214. Mengshoel, O.J., Wilkins, D.C.: Abstraction for belief revision: Using a genetic algorithm to compute the most probable explanation. In: Satisficing Models: Papers from the 1998 AAAI Spring Symposium, pp. 46–53. AAAI Press, Menlo Park (1998) Techical Report SS-98-05

215. Mignotte, M., Collet, C., Pérez, P., Bouthemy, P.: Markov Random Field and Fuzzy Logic Modeling in Sonar Imagery: Application to the Classification of Underwater Floor. Computer Vision and Image Understanding 79(1), 4–24 (2000)

216. Mihalkova, L., Mooney, R.J.: Bottom-up learning of markov logic network structure. In: Proceedings of the Twenty-Fourth International Conference (ICML 2007), pp. 625–632 (2007)

217. Mitchell, T.M.: Machine Learning. McGraw-Hill, New York (1997)

218. Miyahara, K., Pazzani, M.J.: Collaborative Filtering with the Simple Bayesian Classifier. In: Mizoguchi, R., Slaney, J.K. (eds.) PRICAI 2000. LNCS, vol. 1886, pp. 679–689. Springer, Heidelberg (2000)

219. Mood, A.M., Graybill, F.A., Boes, D.C.: Introduction to the Theory of Statistics, 3rd edn. McGraw-Hill, New York (1974)

220. Moody, J., Darken, C.J.: Fast Learning in Networks of Locally-Tuned Processing Units. Neural Computation 1(2), 281–294 (1989)

221. Moussouris, J.: Gibbs and Markov Random Systems with Constraints. Journal of Statistical Physics 10, 11–33 (1974)

222. Muggleton, S.H.: Learning Structure and Parameters of Stochastic Logic Programs. In: Proceedings of the 12th International Conference on Inductive Logic Programming, pp. 198–206. Springer, Heidelberg (2002)

223. Nadaraya, E.A.: On Estimating Regression. Theory of Probability and its Applications 9, 141–142 (1964)

224. Nagel, E.: The Structure of Science. Harcourt, Brace & World, New York (1961)

225. Neapolitan, R.E.: Probabilistic Reasoning in Expert Systems: Theory and Algorithms. John Wiley & Sons, New York (1990)

226. Neapolitan, R.E.: Learning Bayesian Networks. Prentice Hall, Upper Saddle River (2004)

227. Neurath, O.: Protokollsätze. Erkenntnis 3, 204–214 (1932)

228. Neville, J., Jensen, D.: Relational Dependency Networks. Journal of Machine Learning Research 8, 653–692 (2007)

229. Newell, A., Simon, H.A.: Computer Science as Empirical Inquiry: Symbols and Search. Communications of the ACM 19, 113–126 (1976)

230. Newman, D., Hettich, S., Blake, C., Merz, C.: UCI Repository of Machine Learning Databases (1998),
http://www.ics.uci.edu/~mlearn/MLRepository.html

231. Ninio, M., Privman, E., Pupko, T., Friedman, N.: Phylogeny reconstruction: Increasing the accuracy of pairwise distance estimation using bayesian inference of evolutionary rates. Bioinformatics 23(2), 136–141 (2007)

232. Oberhoff, D., Kolesnik, M.: Unsupervised bayesian network learning for object recognition in image sequences. In: Kůrková, V., Neruda, R., Koutník, J. (eds.) ICANN 2008, Part I. LNCS, vol. 5163, pp. 235–244. Springer, Heidelberg (2008)

233. Papoulis, A.: Probability, Random Variables, and Stochastic Processes, 3rd edn. McGraw-Hill, New York (1991)

234. Parzen, E.: On Estimation of a Probability Density Function and Mode. Annals of Mathematical Statistics 33, 1065–1076 (1962)

235. Pearl, J.: Bayesian networks: A model of self-activated memory for evidential reasoning. In: Proceedings of the 7th Conference of the Cognitive Science Society, pp. 329–334 (1985)

236. Pearl, J.: Evidential Reasoning Using Stochastic Simulation of Causal Models. Artificial Intelligence 32, 245–257 (1987)

237. Pearl, J.: Probabilistic Reasoning in Intelligent Systems. Morgan Kaufmann, San Francisco (1988)

238. Pearl, J., Geiger, D., Verma, T.: The Logic of Influence Diagrams. In: Oliver, R., Smith, J. (eds.) Influence Diagrams, Belief Networks and Decision Analysis. Wiley Ltd., Sussex (1990)

239. Pearlmutter, B.A.: Learning state space trajectories in recurrent neural networks. Neural Computation 1, 263–269 (1989)

240. Pérez, P.: Markov Random Fields and Images. CWI Quarterly 11, 413–437 (1998)

241. Pernkopf, F., Bilmes, J.A.: Efficient Heuristics for Discriminative Structure Learning of Bayesian Network Classifiers. Journal of Machine Learning Research 11, 2323–2360 (2010)

242. Pfeffer, A., Koller, D.: Semantics and inference for recursive probability models. In: Proceedings of the Seventeenth National Conference on Artificial Intelligence (AAAI 2000), pp. 538–544. AAAI Press, Menlo Park (2000)

243. Pfeifer, R., Scheier, C.: Understanding Intelligence. MIT Press, Cambridge (1999)

244. Pham, T.V., Worring, M., Smeulders, A.W.M.: Face detection by aggregated bayesian network classifiers. Pattern Recognition Letters 23, 451–461 (2001)

245. Pollastro, P., Rampone, S.: Homo Sapiens Splice Sites Dataset (2003), http://www.sci.unisannio.it/docenti/rampone/

246. Popper, K.R.: The Logic of Scientific Discovery. Hutchinson & Co., London (1959)

247. Quine, W.v.O.: Two Dogmas of Empiricism. The Philosophical Review 60, 20–43 (1951)

248. Quine, W.v.O.: Pursuit of Truth, 2nd edn. Harvard University Press, Cambirdge (1992)

249. Quinlan, J.R.: C4.5 Programs for Machine Learning. Morgan Kaufmann, San Mateo (1993)

250. Quinlan, J.R.: Learning First-Order Definitions of Functions. Journal of Artificial Intelligence Research 5, 139–161 (1996)

251. Rabiner, L.R.: A tutorial on hidden markov models and selected applications in speech recognition. Proceedings of the IEEE 77(2), 257–286 (1989)

252. Rabiner, L.R., Gold, B.: Theory and Application of Digital Signal Processing. Prentice-Hall, Englewood Cliffs (1975)

253. Ravikumar, P., Raskutti, G., Wainwright, M., Yu, B.: Model Selection in Gaussian Graphical Models: High-Dimensional Consistency of ℓ_1-regularized MLE. In: Advances in Neural Information Processing Systems, pp. 1329–1336 (2008)

254. Rennie, J.D., Shih, L., Teevan, J., Karger, D.R.: Tackling the poor assumptions of naive bayes text classifiers. In: Proceedings of the Twentieth International Conference on Machine Learning (ICML 2003), pp. 616–623 (2003)

255. Richard, M.D., Lippmann, R.P.: Neural network classifiers estimate Bayesian a posteriori probabilities. Neural Computation 3, 461–483 (1991)

256. Richardson, M., Domingos, P.: Markov Logic Networks. Machine Learning 62, 107–136 (2006)

257. Richardson, T., Spirtes, P.: Ancestral graph Markov models. Annals of Statistics 30(4), 962–1030 (2002)

258. Ripley, B.D.: Pattern Recognition and Neural Networks. Cambridge University Press, Cambridge (1996)

259. Rissanen, J.: Stochastic Complexity. Journal of the Royal Statistical Society Series B 49, 223–239 (1987)

260. Robinson, R.W.: Counting Unlabeled Acyclic Digraphs. In: Little, C.H.C. (ed.) Combinatorial Mathematics. LNM, vol. 622, pp. 28–43. Springer, New York (1977)
261. Robles, V., Larrañaga, P., Menasalvas, E., Pérez, M.S., Herves, V.: Improvement of Naïve Bayes Collaborative Filtering Using Interval Estimation. In: Proceedings of the IEEE/WIC International Conference on Web Intelligence (WI 2003), pp. 168–174 (2003)
262. Rorty, R.M. (ed.): The Linguistic Turn. Essays in Philosophical Method, 2nd edn. The University of Chicago Press, Chicago (1992) (Original work published 1967)
263. Rosenblatt, M.: Conditional Probability Density and Regression Estimators. In: Krishnaiah, P. (ed.) Multivariate Analysis, vol. II, pp. 25–31. Academic Press, New York (1969)
264. Roweis, S.T., Ghahramani, Z.: A Unifying Review of Linear Gaussian Models. Neural Computation 11(2), 305–345 (1999)
265. Rozanov, Y.A.: Markov Random Fields. Springer, New York (1982)
266. Rue, H.: Fast Sampling of Gaussian Markov Random Fields. Journal of the Royal Statistical Society Series B 63, 325–338 (2001)
267. Ruiz, F.J., Angulo, C., Agell, N.: IDD: A Supervised Interval Distance-Based Method for Discretization. IEEE Transactions on Knowledge and Data Engineering 20, 1230–1238 (2008)
268. Rumelhart, D.E., Hinton, G.E., Williams, R.J.: Learning Internal Representations by Error Propagation, pp. 318–362. MIT Press, Cambridge (1986)
269. Russell, S., Norvig, P.: Artificial Intelligence, 2nd edn. Prentice Hall, Upper Saddle River (2003)
270. Santhanam, N., Wainwright, M.J.: Information-theoretic limits of graphical model selection in high dimensions. In: IEEE International Symposium on Information Theory (ISIT 2008), pp. 2136–2140 (2008)
271. Schwarz, G.: Estimating the Dimension of a Model. The Annals of Statistics 6, 461–464 (1978)
272. Scilla, A.: La vana speculazione disingannata dal senso. Giunti, Firenze (1996) (Original work published 1670)
273. Searle, J.R.: Minds, Brains, and Programs. Behavioral and Brain Sciences 3, 417–424 (1980)
274. Seiler, C., Büchler, P., Nolte, L.P., Paulsen, R., Reyes, M.: Hierarchical Markov Random Fields Applied to Model Soft Tissue Deformations on Graphics Hardware. In: Magnenat-Thalmann, N., Zhang, J.J., Feng, D.D. (eds.) Recent Advances in the 3D Physiological Human, ch.9, pp. 133–148. Springer, London (2009)
275. Sellars, W.: Philosophy and the Scientific Image of Man. In: Colodny, R. (ed.) Frontiers of Science and Philosophy, pp. 35–78. University of Pittsburgh Press, Pittsburgh (1962), http://www.ditext.com/sellars/psim.html
276. Shannon, C.E., Weaver, W.: The Mathematical Theory of Communication. University of Illinois Press, Urbana (1949)
277. Silva, R., Ghahramani, Z.: The Hidden Life of Latent Variables: Bayesian Learning with Mixed Graph Models. Journal of Machine Learning Research 10, 1187–1238 (2009)
278. Silverman, B.: Density Estimation for Statistics and Data Analysis. Chapman and Hall, Boca Raton (1986)
279. Simon, H.A.: Does Scientific Discovery Have a Logic? Philosophy of Science 40, 471–480 (1973)
280. Singla, P., Domingos, P.: Markov Logic in Infinite Domains. In: Proceedings of the Twenty-Third Conference on Uncertainty in Artificial Intelligence, pp. 368–375 (2007)

281. Smith, G.B.: Preface to Stuart Geman and Donald Geman, "Stochastic Relaxation, Gibbs Distributions, and the Bayesian Restoration of Images. In: Fischler, M.A., Firschein, O. (eds.) Readings in Computer Vision: Issues, Problems, Principles, and Paradigms, pp. 562–563. Morgan Kaufmann, Los Altos (1987)
282. Smyth, P.: Belief Networks, hidden Markov Models, and Markov Random Fields: A Unifying View. Pattern Recognition Letters 18, 1261–1268 (1997)
283. Snyman, J.A.: Practical Mathematical Optimization: An Introduction to Basic Optimization Theory and Classical and New Gradient-Based Algorithms. Springer, New York (2005)
284. Specht, D.F.: Probabilistic neural networks. Neural Networks 3(1), 109–118 (1990)
285. Sperduti, A., Starita, A.: Supervised neural networks for the classification of structures. IEEE Transactions on Neural Networks 8(3), 714–735 (1997)
286. Spirtes, P., Glymour, C., Scheines, R.: Causation, Prediction, and Search, 2nd edn. MIT Press, Cambridge (2001) (Original work published 1993 by Springer-Verlag)
287. Srinivas, S.: A Probabilistic Approach to Hierarchical Model-based Diagnosis. In: Proceedings of the Tenth Annual Conference on Uncertainty in Artificial Intelligence (UAI 1994), pp. 538–545 (1994)
288. Stainvas, I., Lowe, D.: A Generative Model for Separating Illumination and Reflectance from Images. Journal of Machine Learning Research 4, 1499–1519 (2003)
289. Stegmüller, W.: The Structuralist View of Theories. Springer, New York (1979)
290. Stenger, V.J.: God and rev. bayes. Skeptical Briefs 17(2) (2007)
291. Stich, S.P.: The Fragmentation of Reason: Preface to a Pragmatic Theory of Cognitive Evaluation. MIT Press, Cambridge (1990)
292. Stich, S.P.: Naturalizing Epistemology: Quine, Simon and the Prospects for Pragmatism. In: Hookway, C., Peterson, D. (eds.) Philosophy and Cognitive Science. 40, pp. 1–17. Cambridge University Press, Cambridge (1993)
293. Strawson, P.F. (ed.): Analysis and Metaphysics. An Introduction to Philosophy. Oxford University Press, New York (1992)
294. Sutton, C., McCallum, A.: An introduction to conditional random fields for relational learning. In: Getoor and Taskar [109]
295. Taskar, B., Guestrin, C., Koller, D.: Max-Margin Markov Networks. In: Advances in Neural Information Processing Systems, NIPS 2003 (2004)
296. Torrione, P.A., Collins, L.: Application of Markov random fields to landmine detection in ground penetrating radar data. In: Harmon, R.S., Holloway, J.H.J., Broach, J.T. (eds.) Detection and Sensing of Mines, Explosive Objects, and Obscured Targets XIII, vol. 6953 (2008)
297. Trentin, E.: Simple and Effective Connectionist Nonparametric Estimation of Probability Density Functions. In: Schwenker, F., Marinai, S. (eds.) ANNPR 2006. LNCS (LNAI), vol. 4087, pp. 1–10. Springer, Heidelberg (2006)
298. Trentin, E., Di Iorio, E.: Classification of graphical data made easy. Neurocomputing 73(1-3), 204–212 (2009)
299. Trentin, E., Freno, A.: Probabilistic Interpretation of Neural Networks for the Classification of Vectors, Sequences, and Graphs. In: Bianchini, M., Maggini, M., Scarselli, F., Jain, L.C. (eds.) Innovations in Neural Information Paradigms and Applications. SCI, vol. 247, pp. 155–182. Springer, Heidelberg (2009)
300. Trentin, E., Freno, A.: Unsupervised Nonparametric Density Estimation: A Neural Network Approach. In: Proceedings of the International Joint Conference on Neural Networks (IJCNN 2009), pp. 3140–3147. IEEE, Los Alamitos (2009)
301. Trentin, E., Gori, M.: A survey of hybrid ANN/HMM models for automatic speech recognition. Neurocomputing 37, 91–126 (2001)

302. Trentin, E., Gori, M.: Robust Combination of Neural Networks and Hidden Markov Models for Speech Recognition. IEEE Transactions on Neural Networks 14(6), 1519–1531 (2003)
303. Trentin, E., Rigutini, L.: A maximum-likelihood connectionist model for unsupervised learning over graphical domains. In: Alippi, C., Polycarpou, M., Panayiotou, C., Ellinas, G. (eds.) ICANN 2009. LNCS, vol. 5768, pp. 40–49. Springer, Heidelberg (2009)
304. Trianni, G., Gamba, P.: Fast damage mapping in case of earthquakes using multitemporal SAR data. Journal of Real-Time Image Processing 4, 195–203 (2009)
305. Tsamardinos, I., Aliferis, C.F., Statnikov, A.R.: Time and Sample Efficient Discovery of Markov Blankets and Direct Causal Relations. In: KDD 2003: Proceedings of the Ninth ACM SIGKDD International Conference on Knowledge Discovery and Data Mining, pp. 673–678. ACM, New York (2003)
306. Turing, A.M.: Computing Machinery and Intelligence. Mind 59, 433–460 (1950)
307. Unwin, S.D.: The Probability of God. Three Rivers Press, New York (2003)
308. Uwents, W., Blockeel, H.: Classifying relational data with neural networks. In: Kramer, S., Pfahringer, B. (eds.) ILP 2005. LNCS (LNAI), vol. 3625, pp. 384–396. Springer, Heidelberg (2005)
309. Uwents, W., Monfardini, G., Blockeel, H., Scarselli, F., Gori, M.: Two connectionist models for graph processing: An experimental comparison on relational data. In: MLG 2006, Proceedings of the International Workshop on Mining and Learning with Graphs, pp. 211–220 (2006)
310. Vapnik, V.N.: The Nature of Statistical Learning Theory. Springer, New York (1995)
311. Watson, G.S.: Smooth Regression Analysis. Sankhyā: The Indian Journal of Statistics, Series A 26, 359–372 (1964)
312. Webb, G.I., Boughton, J.R., Wang, Z.: Not so naive bayes: Aggregating one-dependence estimators. Machine Learning 58, 5–24 (2005)
313. Wellman, M.P., Liu, C.L.: State-Space Abstraction for Anytime Evaluation of Probabilistic Networks. In: Proceedings of the Tenth Conference on Uncertainty in Artificial Intelligence, pp. 567–574. Morgan Kaufmann, Seattle (1994)
314. Wermuth, N., Lauritzen, S.L.: On Substantive Research Hypotheses, Conditional Independence Graphs and Graphical Chain Models. Journal of the Royal Statistical Society Series B 52, 21–50 (1990)
315. Whittaker, J.: Graphical Models in Applied Multivariate Statistics. John Wiley & Sons, Chichester (1990)
316. Wigmore, J.H.: The Problem of Proof. Illinois Law Review 8(2), 77–103 (1913)
317. Witten, I.H., Frank, E.: Data Mining, 2nd edn. Morgan Kaufmann, San Francisco (2005)
318. Wright, S.: Correlation and Causation. Journal of Agricultural Research 20, 557–585 (1921)
319. Wu, J., Chung, A.C.S.: A segmentation model using compound Markov random fields based on a boundary model. IEEE Transactions on Image Processing 16(1), 241–252 (2007)
320. Yang, F., Jiang, T.: Pixon-based image segmentation with Markov random fields. IEEE Transactions on Image Processing 12(12), 1552–1559 (2003)
321. Yang, Y., Webb, G.I.: On why discretization works for naive-bayes classifiers. In: 16th Australian Conference on Artificial Intelligence, pp. 440–452 (2003)
322. Yedidia, J.S., Freeman, W.T., Weiss, Y.: Generalized Belief Propagation. In: Advances in Neural Information Processing Systems, pp. 689–695 (2000)
323. Yoshida, R., West, M.: Bayesian Learning in Sparse Graphical Factor Models via Variational Mean-Field Annealing. Journal of Machine Learning Research 11, 1771–1798 (2010)

324. Zheng, H., Daoudi, M., Jedynak, B.: Blocking adult images based on statistical skin detection. Electronic Letters on Computer Vision and Image Analysis 4(2), 1–14 (2004)
325. Zheng, Y., Li, H., Doermann, D.: Machine Printed Text and Handwriting Identification in Noisy Document Images. IEEE Transactions on Pattern Analysis and Machine Intelligence 26, 337–353 (2004)
326. Zhou, S., Lafferty, J.D., Wasserman, L.A.: Time Varying Undirected Graphs. Machine Learning 80, 295–319 (2010)
327. Zhu, J., Nie, Z., Zhang, B., Wen, J.R.: Dynamic Hierarchical Markov Random Fields for Integrated Web Data Extraction. Journal of Machine Learning Research 9, 1583–1614 (2008)
328. Zhu, X.: Semi-Supervised Learning Literature Survey. Tech. Rep. 1530, Computer Sciences, University of Wisconsin – Madison (2005),
    http://pages.cs.wisc.edu/~jerryzhu/pub/ssl_survey.pdf

Index